body structures and functions

5th edition
body structures and functions

ELVIRA B. FERRIS B.A., M.S.
ESTHER G. SKELLEY R.N., M.S.

DELMAR PUBLISHERS
COPYRIGHT ©1979
BY LITTON EDUCATIONAL PUBLISHING, INC.

All rights reserved. Certain portions of this work copyright ©1954, 1964, 1967, 1973 by Litton Educational Publishing, Inc. No part of this work covered by the copyright hereon may be reproduced or used in any form or by any means — graphic, electronic, or mechanical, including photocopying, recording, taping, or information storage and retrieval systems — without written permission of the publisher.

10 9 8 7 6 5 4 3
LIBRARY OF CONGRESS CATALOG CARD NUMBER: 77-83347
ISBN: 0-8273-1322-5

Printed in the United States of America
Published Simultaneously in Canada by
Delmar Publishers, A Division of
Van Nostrand Reinhold, Ltd.

Technical Consultant
Ruth Helen Adams, Ed.D.

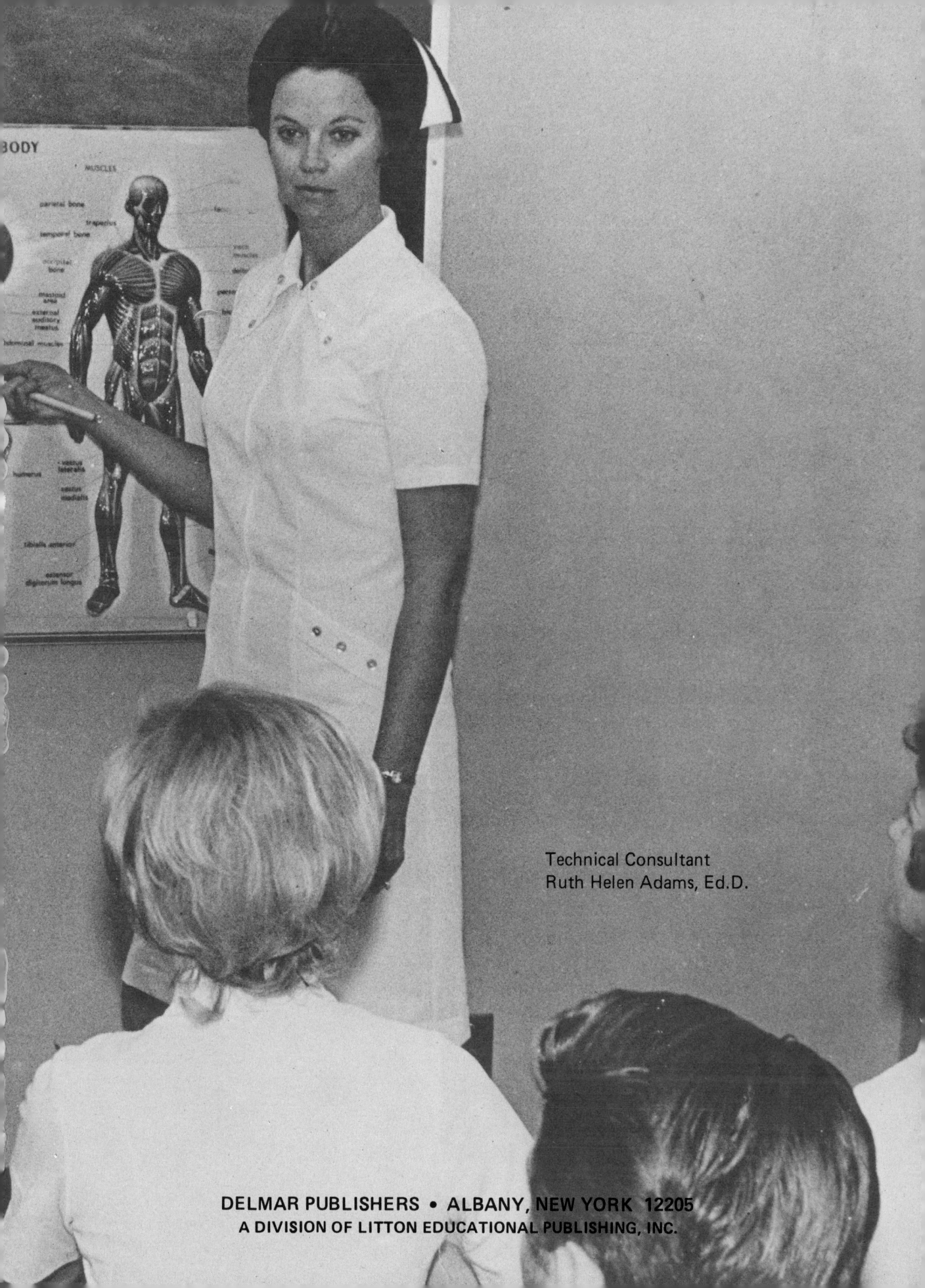

DELMAR PUBLISHERS • ALBANY, NEW YORK 12205
A DIVISION OF LITTON EDUCATIONAL PUBLISHING, INC.

PREFACE

Body Structures and Functions, now in its fifth edition, presents an overview of each body system and an examination of the integrated functions of those systems. The text helps the student study the human body by focusing on the essential components and processes within each system. Tables and diagrams organize technical terms and concepts to facilitate comprehension. Illustrations of detailed body structures are accompanied by explanations of their functions.

Body Structures and Functions is particularly adapted for use by practical nurses and other persons working in allied health fields, who require an understanding of structures and processes. Specific behavioral objectives help the student to define goals for each unit of study. Suggested activities present thought-provoking topics for discussion and when appropriate, guidelines for laboratory study. Reviews provide an opportunity for evaluation at the end of the units. Sectional self-evaluation exercises summarize the material presented on each body system.

Changes made in this revision primarily focus on the unit-end evaluations. Questions have been modified to present a greater variety and to give a more comprehensive review of the unit material. Other changes in the suggested activities and portions of the text reflect the continuing efforts to incorporate technological change into educational materials.

Elvira B. Ferris obtained a bachelor's degree in Biology from Hunter College and later earned a master's degree in Education from City College of New York. Ms. Ferris has also done extensive graduate work in psychology and guidance. Her previous publications include another Delmar text, *Microbiology for Health Careers.*

Esther G. Skelley is a registered nurse and received her undergraduate and graduate degrees in Public Health from Columbia University in New York. She is the author of *Medications and Mathematics for the Nurse.*

Other Delmar texts in the Practical Nursing Series are:

>Nutrition and Diet Modifications – C. Townsend
>Emotional Adjustment to Illness – K. Noonan
>Understanding Human Behavior – M. Milliken
>Medical-Surgical Nursing Procedures – L. Broadwell and B. Milutinovic

ACKNOWLEDGMENTS

Staff at Delmar Publishers
>Health Occupations Division
>>Supervising Editor: Angela R. Emmi, RN
>>Source Editor: Anne Greatbatch, RN
>>Editorial Assistant: Hazel Kozakiewicz

CONTENTS

SECTION 1 THE BODY AS A WHOLE

Unit 1	Introduction to the Structural Units	1
Unit 2	Cells	4
Unit 3	Tissues	7
Unit 4	Organs and Systems	10
	Self-Evaluation	12

SECTION 2 THE BODY FRAMEWORK

Unit 5	Introduction to the Skeletal System	14
Unit 6	Structure and Formation of Bone	18
Unit 7	Parts of the Skeleton	21
Unit 8	Injuries and Diseases of Bones and Joints	25
	Self-Evaluation	27

SECTION 3 BODY MOVEMENT

Unit 9	Introduction to the Muscular System	28
Unit 10	Attachment of Muscles	30
Unit 11	Principal Skeletal Muscles	32
Unit 12	Abnormal Muscular Conditions	35
	Self-Evaluation	37

SECTION 4 TRANSPORT OF FOOD AND OXYGEN

Unit 13	Introduction to the Circulatory System	38
Unit 14	The Heart	41
Unit 15	Function and Path of General Circulation	45
Unit 16	Pulmonary Circulation	49
Unit 17	Blood Vessels	51
Unit 18	The Blood	54
Unit 19	Lymphatic System	59
Unit 20	Disorders of the Circulatory System	62
	Self-Evaluation	65

SECTION 5 BREATHING PROCESSES

Unit 21	Introduction to the Respiratory System	67
Unit 22	Respiratory Organs and Structures	69
Unit 23	Mechanics of Breathing	72
Unit 24	Respiratory Disorders	74
	Self-Evaluation	77

SECTION 6 DIGESTION OF FOOD

Unit 25	Introduction to the Digestive System	79
Unit 26	Digestion in the Stomach	83
Unit 27	Digestion in the Small Intestine	86
Unit 28	The Large Intestine	89
Unit 29	Disorders of the Digestive System	91
	Self-Evaluation	93

SECTION 7 ELIMINATION OF WASTE MATERIALS

Unit 30	Introduction to the Excretory System	95
Unit 31	Urinary System	97
Unit 32	The Skin	101
Unit 33	Disorders of the Excretory System	103
	Self-Evaluation	106

SECTION 8 HUMAN REPRODUCTION

Unit 34	Introduction to the Reproductive System	108
Unit 35	The Organs of Reproduction	110
Unit 36	Disorders of the Reproductive System	115
	Self-Evaluation	117

SECTION 9 REGULATORS OF BODY FUNCTIONS

Unit 37	Introduction to the Endocrine System	118
Unit 38	Pituitary Gland	120
Unit 39	The Thyroid and Parathyroid Glands	123
Unit 40	Functions of Adrenal Glands and Gonads	125
Unit 41	The Pancreas	127
Unit 42	Endocrine Gland Disorders	129
	Self-Evaluation	132

SECTION 10 COORDINATION OF BODY FUNCTIONS

Unit 43	Introduction to the Nervous System	133
Unit 44	The Central Nervous System: Brain and Spinal Cord	135
Unit 45	The Peripheral and Autonomic Nervous Systems	138
Unit 46	Special Sense Organs — Eye and Ear	141
Unit 47	Diseases of the Nervous System	145
	Self-Evaluation	148

SECTION 11 CARE OF THE BODY SYSTEMS

Unit 48	Practices for Health Maintenance	150

Glossary	155
Metric System Equivalents	160
Index	162

SECTION 1

THE BODY AS A WHOLE

unit 1 introduction to the structural units

OBJECTIVES

Studying this unit should help the student:

- Identify the life processes.
- Define the chemical processes which maintain life.
- Identify the body cavities and the organs they contain.

A man is born, he grows, he matures, and he dies. In the intervening years between his birth and death, his body carries on the several life functions which keep him alive and usually active. As is true of all living things, he has inherited a certain size range, form, and lifespan from his forebears. Man inherits these many characteristics through the sperm and egg from his parents. The creation of this living condition depends for its very existence on the constant release of energy in every cell of his body. Powered by the energy released from food, the cells are able to maintain their own living condition and thus the life of a human being.

The simplest forms of life consist of a relatively few cells; all of the cells carry on all of the life functions: breathing; eating, digesting, absorbing and assimilating food; secreting important digestive juices, enzymes, and hormones; excreting wastes; moving and adapting to environment; and reproducing. However, an extremely complex form such as man consists of over fifty thousand billion cells, and early in human development certain groups of cells become highly specialized for certain functions (such as motion or response). These special cell groups are called *tissues*. They, in turn, are grouped into larger functioning structures known as *organs*.

The functional activities of cells that result in growth, repair, release of energy, use of food, and secretion are combined under the heading of *metabolism*. It consists of two opposite processes: (1) the building up of complex materials from simpler ones, called *anabolism*, and (2) the tearing down and changing of more complex substances into simpler ones, with a release of energy, called *catabolism*. Thus, the sum of all the chemical reactions within a cell is metabolism.

Organ systems are made up of the several organs concerned with a special function in the body. The digestive system, for example, is composed of all organs involved in digestion. The circulatory system includes all organs involved in circulation.

The organs making up most of the nine systems of man are distributed in the dorsal and ventral cavities. The dorsal cavity contains the brain and spinal cord. The dorsal cavity may also be referred to as two separate cavities: the brain is located in the *cranial* cavity; the spinal cord, in the *spinal* cavity. The *diaphragm* divides the ventral cavity into

1

THE BODY AS A WHOLE

two parts: the upper thoracic and lower abdominopelvic cavities. The *thoracic* cavity contains the heart and lungs. The *abdominopelvic* cavity is really one large cavity with no separation between the two cavities. However, to avoid confusion, it is referred to separately as the abdominal and pelvic cavities, figure 1-1. The abdominal cavity contains the stomach, liver, gallbladder, pancreas, spleen, small intestine, appendix, and part of the large intestine, see figure 1-2. The kidneys are in the back of the abdominal cavity under its lining. The urinary bladder, the reproductive organs, and the remainder of the large colon are in the pelvic area.

The functioning of a complex organism such as a human being includes more than the activities of cells and tissues. Organs and organ systems are also vitally concerned.

The remainder of this section is devoted, first of all, to the basic unit of structure and function, the cell; secondly, to the specialized groups of cells, the tissues; and finally to the organs of the various systems. In the other sections, each of the body systems is discussed in detail.

Fig. 1-1 Cavities of the body

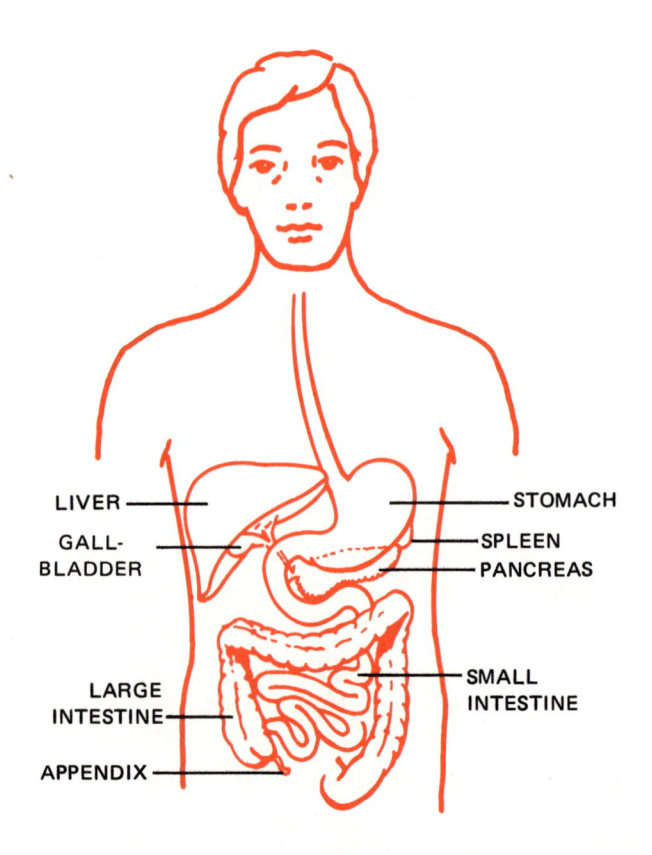

Fig. 1-2 Abdominal cavity

INTRODUCTION TO THE STRUCTURAL UNITS

SUGGESTED ACTIVITIES

- Look up and learn the meaning, spelling, and proper use of the following terms.

anatomy	cranium	therapeutic medicine	organism
physiology	interior	preventive medicine	thorax
hygiene	anterior	lower extremities	dorsal
excretory	posterior	upper extremities	ventral

REVIEW

Match each of the terms in column I with its correct statement in column II.

Column I	Column II
_____ 1. catabolism	a. the sum of all the chemical reactions within the cell
_____ 2. pelvic cavity	b. constructive chemical processes which use food to build complex materials of the body
_____ 3. cranial cavity	
_____ 4. anabolism	
_____ 5. abdominal cavity	c. useful breakdown of food materials resulting in the release of energy
_____ 6. dorsal cavity	d. contained within the thoracic cavity
_____ 7. metabolism	e. cavity in which the reproductive organs, urinary bladder, and lower part of large intestine are located
_____ 8. tissue	
_____ 9. kidneys, ureters and adrenal glands	f. cavity in which the stomach, liver, gallbladder, pancreas, spleen, appendix, cecum, and colon are located
_____ 10. heart and lungs	g. the cavity containing both the cranial and spinal cavities
	h. a group of cells which together perform a particular job
	i. portion of the dorsal cavity containing the brain
	j. divides the ventral cavity into two regions
	k. structures located behind the abdominal cavity and under its lining
	l. the abdominopelvic cavity

unit 2 cells

OBJECTIVES

Studying this unit should help the student:

- Identify the structure of a typical cell.
- Define the function of each main component of a typical cell.
- Relate the function of cells to the function of the body.

When a meadow of grass is seen from a distance, it appears to be a solid green carpet spread upon the earth; however, closer observation shows that this green carpet is not a solid mass, but is made up of many separate blades of grass. So it is with the body of a plant or animal; it seems to be one whole, but when a portion of it is examined under a microscope, it is found to be made up of many discrete parts. These small parts, or units, are called *cells*, which make up all plants and all animals; man is no exception.

Most cells are so small they cannot be seen with the naked eye. If, however, a very thin tissue slice of liver, skin, or some other part of the body is placed under the microscope, the cells of that tissue can be seen easily. Each cell appears to be made of two parts — a dense inner portion called the *nucleus*, and a lighter outer portion called the *cytoplasm*. The exposed outer edge of the cytoplasm forms a thin layer known as the *cell membrane*. The cell membrane separates the cell from neighboring cells and from the fluids that bathe it. It also regulates the flow of substances into and out of the cell by allowing certain molecules to pass through while preventing the passage of others.

Under a regular microscope, the cytoplasm itself looks to be a semifluid material. The electron microscope has shown greater detail and provided new knowledge of the structure of cells. Electron microscope photographs reveal this seemingly structureless cytoplasm to be a network of fine membranes enclosing tubules, swollen areas, and flattened sacs, all with many interconnections. Some of the membranes are covered with minute granules now known to be the sites of protein manufacture. Other membranes are smooth and appear to function in transport and in manufacturing processes (chemical syntheses).

In the cytoplasm, also, are larger structures just visible under the light microscope. These are the *mitochondria*, the powerhouses of the cell, where energy is released from food and sent to where it is needed.

The nucleus serves two purposes. It is the control center for all cell activity and also governs inheritance. The largest structure in the nucleus is the *nucleolus* which stores proteins. Although most cells contain only a single nucleus, certain large cells, such as the muscle cells, contain several nuclei.

There are many kinds of cells of different shapes and sizes. Most of them have the characteristics shown in figure 2-1. This is only a generalized diagram of a very simple cell, however. Some of the more specialized types, such as nerve cells and red blood cells, look very different, figure 2-2.

When a cell reaches a certain size it may divide to form two daughter cells. When this happens, the nucleus divides first by a process called *mitosis*. During this process the nuclear

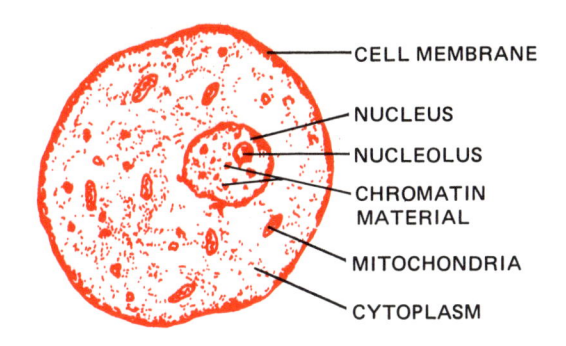

Fig. 2-1 A simple cell

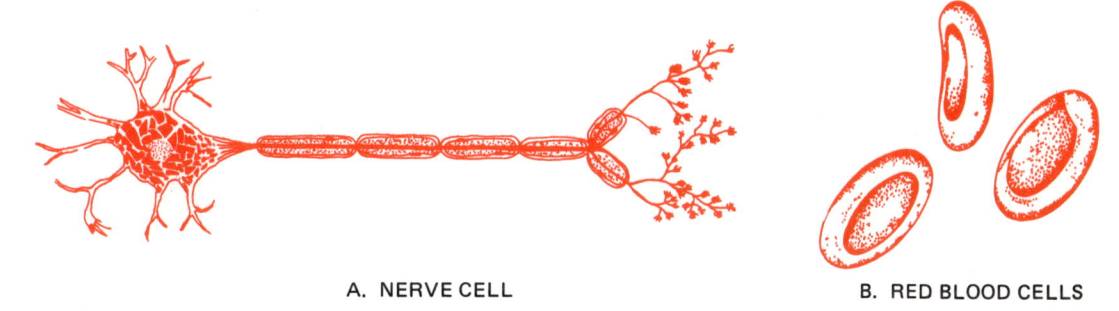

Fig. 2-2 Specialized cells
A. NERVE CELL
B. RED BLOOD CELLS

material is distributed to each of the two daughter nuclei. This is followed by the division of the cytoplasm into two approximately equal parts by the formation of a new membrane between the two nuclei. It is only during the process of nuclear division that structures called *chromosomes* can be seen. They contain the hereditary material, deoxyribonucleic acid (DNA); it is located in the *chromatin material* in the cell's nucleus. Hereditary information is coded into this basic hereditary material which is found in all cells of living things. DNA also controls certain chemical processes in the cell.

Man is composed entirely of cells and the nonliving substances which cells build up around themselves. The interaction of the various parts of the cell within the cellular structure constitutes the life of the cell. These interactions result in the life activities, life processes, or life functions that were discussed in unit 1. However, in complex organisms, groups of cells become specialists in a particular function. Nerve cells, for example, have become specialized in response; red blood cells, in oxygen transport. Such specialized cells lose the ability to perform some of the other functions, such as reproduction (cell division). Normally, when nerve cells are destroyed or damaged, others cannot be formed to replace them. Specialization also has resulted in an interdependence among cells — cells depending on other kinds of cells to aid them in carrying on the total life activities of the organism. In man, this specialization and interdependence extends to the organs.

SUGGESTED ACTIVITIES

- Scrape the lining of your cheek with a clean tongue depressor. Place these scrapings on a glass slide to which a drop of water has previously

THE BODY AS A WHOLE

been added. Complete the demonstration by adding a thin piece of glass called a cover slip over the drop of water. Observe the slide under the microscope using both low and high power magnifications. Repeat the above demonstration using an iodine stain to accentuate the parts of the cell. Observe and make a diagram of the cells.

- Clean the fingertip with 70 percent alcohol and allow it to dry. Prick the fingertip with a sterilized needle. Smear the blood on a clean glass slide. Observe blood cells under low and high power magnifications. Repeat the above demonstration using a staining solution.

REVIEW

A. Label the typical cell shown.

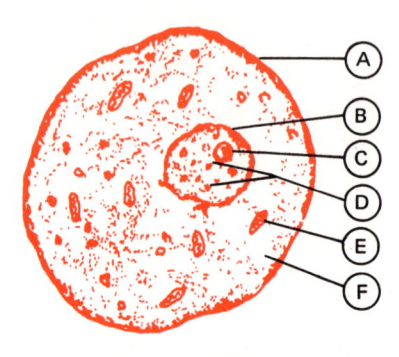

  (A) _____
  (B) _____
  (C) _____
  (D) _____
  (E) _____
  (F) _____

B. Match each of the terms in column I with its correct statement in column II.

Column I	Column II
____ 1. nucleus	a. small units of which all plants and animals are made
____ 2. chromatin material	b. the exposed outer edge of the cell
____ 3. DNA	c. the process by which cells divide
____ 4. cytoplasm	d. an example of a specialized cell
____ 5. nerve	e. the dense inner portion of the cell
____ 6. reproduction	f. the powerhouse in cells from which energy is released
____ 7. cells	g. the light outer portion of the cell
____ 8. mitosis	h. an ability lost by some specialized cells
____ 9. mitochondria	i. the hereditary material within the chromosome
____ 10. cell membrane	j. cell structure where chromosomes are located
	k. cells which specialize in movement

6

unit 3 tissues

OBJECTIVES

Studying this unit should help the student:

- Describe how cells are organized into tissues.
- List the four main types of tissues.
- Define the function and location of tissues.

Specialization of cells results in the differentiation of cells, that is, cells becoming different from each other. This difference shows in the appearance and in the specialized function of the various kinds of cells. These groups of different kinds of specialized cells make up the various tissues. It can be clearly seen under the microscope that the cells making up the various tissues have different shapes. Also seen are the various nonliving substances that the cells build up around themselves. A piece of tissue consists of more than just the cells themselves. This is true especially of the supporting tissues such as bones and cartilage, figures 3-1 and 3-2.

There are four main types of tissue: (1) epithelial, (2) connective, (3) muscle, and (4) nervous. Each has a specialized structure to perform a particular function. The variations, functions, and locations of each type are given in figure 3-3, page 8.

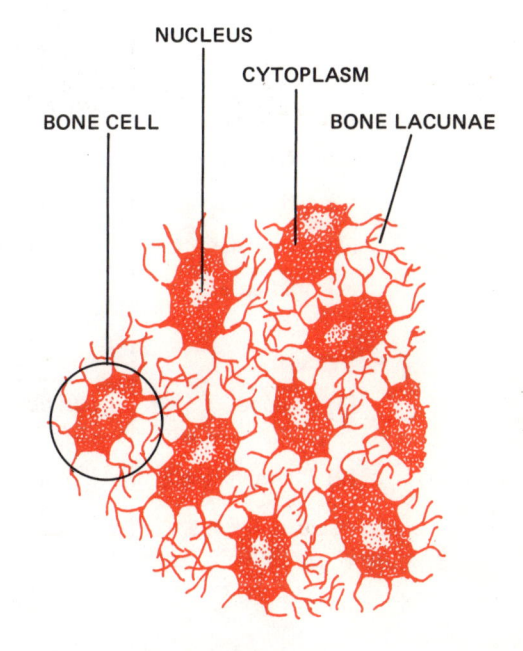

Fig. 3-1 Bone tissue

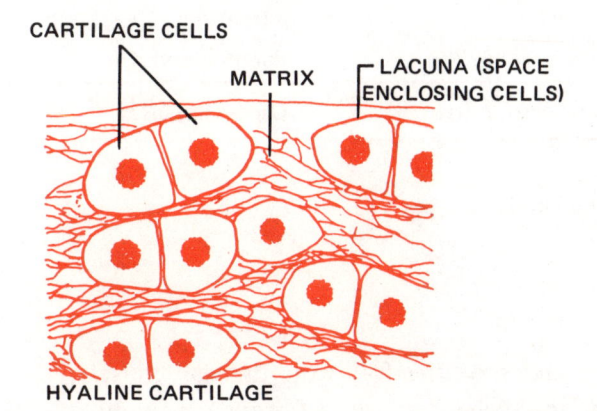

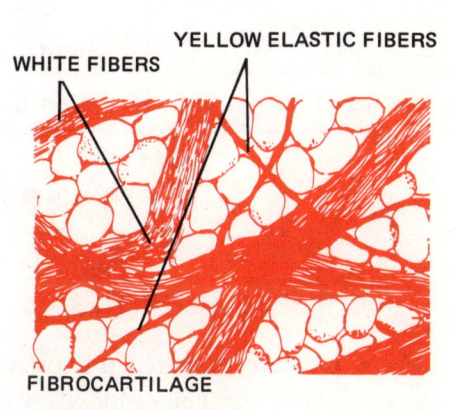

Fig. 3-2 Cartilage tissues

THE BODY AS A WHOLE

TISSUE	LOCATION	FUNCTION
1. Epithelial	Covering surface of body (skin) Lining nose, throat and windpipe, Lining of all digestive tract and many glands	Provides protection; Produces secretions
2. Connective		
a. Bone	Skeleton	Supports and protects
b. Cartilage		
(1) hyaline	Bone surfaces, ribs, nose, larynx, trachea	Supports and protects
(2) fibrocartilage	Intervertebral disks; joints	Supports and protects
c. Dense fibrous	Tendons, ligaments, membranes around bones	Acts as a cushion; joins muscles to bones and bones to bones;
d. Loose fibrous		
(1) fibroelastic	Encases organs; beneath skin	Holds organs together
(2) fibroareolar	Tissue interspaces	Acts as filler tissue
(3) reticular	Tissue interspaces	Acts as filler tissue
(4) adipose	Tissue interspaces	Acts as filler tissue; Cushions and insulates; Stores fat
e. Vascular		
(1) blood	In heart and blood vessels	Transports nutrients and wastes
(2) lymph	Fluid in tissue spaces between cells	Bathes the cells
3. Muscle		
a. Smooth	Walls of many organs	Provides for involuntary movement
b. Skeletal	Attached to bones, tendons, and other muscles	Provides for voluntary movement
c. Cardiac	Heart	Pumps blood
4. Nervous	Brain, Spinal cord, Nerves	Carries impulses

Fig. 3-3 Four main tissue types

SUGGESTED ACTIVITIES

- Observe prepared microscope slides of the following tissues: muscle, nerve, epithelial, blood, bone, cartilage, fat, white fibrous, and yellow elastic.

REVIEW

Match each of the terms in column I with its correct statement in column II.

Column I	Column II
_____ 1. heart	a. provides protection to the body and produces secretions
_____ 2. vascular tissue	b. hardest body tissue providing support
_____ 3. epithelial tissue	c. slightly flexible tissue found in the intervertebral disks and the ribs
_____ 4. cartilage tissue	d. primarily transports nutrients and wastes
_____ 5. differentiation	e. provides for involuntary movements
_____ 6. tissue interspaces	f. provides for voluntary movement
_____ 7. smooth muscle	g. carries impulses and messages throughout the body
_____ 8. bone tissue	h. location of cardiac muscle
_____ 9. nervous tissue	i. result of cell specialization
_____ 10. skeletal muscle	j. location of most loose fibrous connective tissue
	k. fluid in tissue interspaces
	l. cell digestion

unit 4 organs and systems

OBJECTIVES

Studying this unit should help the student:

- Define an organ.
- Define a system.
- Relate various organs to their respective systems.

An organ is made up of tissues which are grouped together to help carry out certain activities. For example, the stomach is an organ which contains nerve tissue, muscular tissue, epithelial tissue, connective tissue, and blood tissue. These tissues function together. The stomach, along with other organs, functions within the digestive system, figure 4-1.

SYSTEM	SYSTEM FUNCTIONS	ORGANS	
Skeletal	Gives shape to body; Protects delicate parts of body; Provides space for attaching muscles; Is instrumental in forming blood; Stores minerals	Skull, Spinal Column, Ribs and Sternum, Shoulder Girdle, Pelvic Girdle, Upper and Lower Extremities	
Muscular	Determines posture; Produces body heat; Provides for movement	Voluntary Muscles (Skeletal) Involuntary Muscles Cardiac Muscle	
Digestive	Prepares food for absorption and use by body cells through modification of chemical and physical states	Mouth, Salivary Glands, Teeth, Pharynx, Esophagus, Stomach, Intestines, Liver, Gallbladder, Pancreas	
Respiratory	Acquires oxygen; Rids body of carbon dioxide	Nose, Larynx, Trachea, Bronchi, Lungs	
Circulatory	Carries oxygen and nourishment to cells of body; Carries waste from cells	Heart, Arteries, Veins, Capillaries, Lymphatic Vessels, Spleen	
Excretory	Removes waste products of metabolism from body	Skin, Lungs, Kidneys, Bladder, Ureters, Urethra	
Nervous	Communicates; Controls body activity; Coordinates body activity	Brain, Nerves, Spinal Cord	
Endocrine	Manufactures hormones to regulate organ activity	Glands (ductless): Pituitary, Thyroid, Parathyroid, Pancreas, Adrenal, Gonads	
Reproductive	Reproduces human beings	Male Testes Scrotum Epididymis Seminal ducts Seminal vesicles Prostate gland Penis Urethra	Female Ovaries Fallopian tubes Uterus Vagina External genitals Breasts

Fig. 4-1 The nine body systems

ORGANS AND SYSTEMS

A group of organs which performs a special function make up a system. Thus, the digestive system has the special function of changing solid food to a liquid. All the organs which participate in any way are part of the digestive system. For example, the digestive system includes the following organs: mouth, salivary glands, stomach, small intestine, liver, pancreas, gallbladder, and large intestine. The circulatory system includes the heart, arteries, veins, capillaries, lymphatic vessels, and spleen.

As each of the systems in the body is studied, it can be seen that each one is adapted to carry on its own special activity. The systems of the body are the skeletal, muscular, digestive, respiratory, circulatory, reproductive, excretory, endocrine, and nervous systems. The functions and organs of each system are shown in figure 4-1.

SUGGESTED ACTIVITIES

- Identify all the types of tissue present in the arm.
- Discuss how each system is involved in the functioning of the arm.

REVIEW

A. Explain the relationship of cells to tissues, organs, and systems.

B. List two organs found in each of the nine body systems.

Skeletal

Muscular

Digestive

Respiratory

Circulatory

Excretory

Nervous

Endocrine

Reproductive

section 1 the body as a whole
SELF-EVALUATION

A. Complete the following statements.

1. A group of similar cells which performs one special function is _____ .

2. Three types of muscle tissue are _____ , _____ , and _____ .

3. Tissue which is found on the surface of the body or lining the body cavities is called _____ .

4. A group of tissues performing one special function is _____ .

5. The part of a cell which directs its activities is _____ .

6. The tissue which provides transportation of materials within the body is _____ .

7. A group of organs which together perform a special function is called _____ .

8. The hardest of the connective tissue adapted to give support and protection is _____ .

9. The tissue which provides for contraction is _____ .

10. Four different kinds of connective tissue are _____ , _____ , _____ , and _____ .

B. Classify each of the following according to its main tissue type.

1. Skin _____
2. Blood _____
3. Adipose _____
4. Heart _____
5. Brain _____

C. Name the body system to which each of the following organs belongs.

1. Brain _____
2. Adrenal glands _____
3. Spinal column _____
4. Voluntary muscles _____
5. Lungs _____
6. Heart _____
7. Kidneys _____
8. Uterus _____
9. Stomach _____

D. Label the parts indicated in the following diagram.

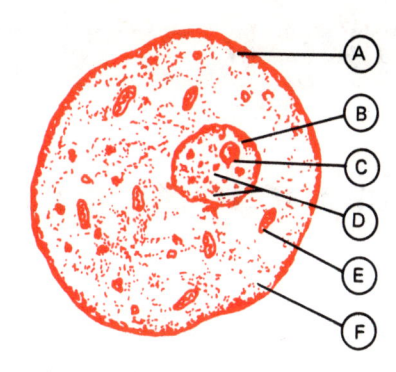

A. _____
B. _____
C. _____
D. _____
E. _____
F. _____

E. Why is the cell called the basic unit of body structure and function?

F. Match each of the terms in column I with its correct statement in column II.

	Column I	Column II
_____	1. abdominal cavity	a. location of brain
_____	2. abdominopelvic cavity	b. another name for chest
_____	3. dorsal cavity	c. location of diaphragm
_____	4. thoracic cavity	d. location of liver
_____	5. ventral cavity	e. location of urinary bladder

SECTION 2

THE BODY FRAMEWORK

unit 5 introduction to the skeletal system

OBJECTIVES

Studying this unit should help the student:

- List the main functions of bones in the body.
- Identify and locate four types of bones.
- Name and define the main types of joints.
- Name the main types of joint motion.

The *skeletal system* comprises the bony framework of the body. It is composed of 206 bones in the adult and performs three main functions: support, protection, and movement. Bones give shape to the body as well as provide it with support. Many provide protection for the soft and delicate organs of the body: for example, the cranium protects the brain, the ribs protect the heart and lungs. Bones provide a place for the attachment of muscles, playing a part in the movement of the body by serving as passively operated levers.

An additional function of bones is providing a storehouse of minerals that the body could draw from in case of inadequate nutrition. The long bones of the body are the site of blood cell formation.

TYPES OF BONES

Bones are classified as one of four types on the basis of their form, figure 5-1. *Long* bones are found in both upper and lower arms and legs. The bones of the skull are examples of *flat* bones — as are the ribs. *Irregular* bones are represented by bones of the spinal column. The wrist and ankle bones are examples of *short* bones, which appear cube-like in shape.

The bones in the hand are short, making flexible movement possible. The same is

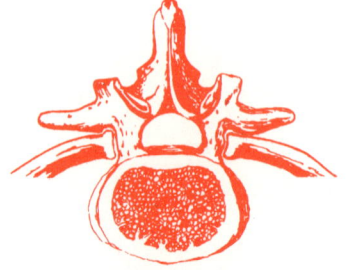

LONG   SHORT   IRREGULAR   FLAT

Fig. 5-1 Bone shapes

INTRODUCTION TO THE SKELETAL SYSTEM

MOVABLE BALL AND SOCKET JOINT (HIP)

PARTIALLY MOVABLE (RIB AND VERTEBRA)

IMMOVABLE (TWO CRANIAL BONES)

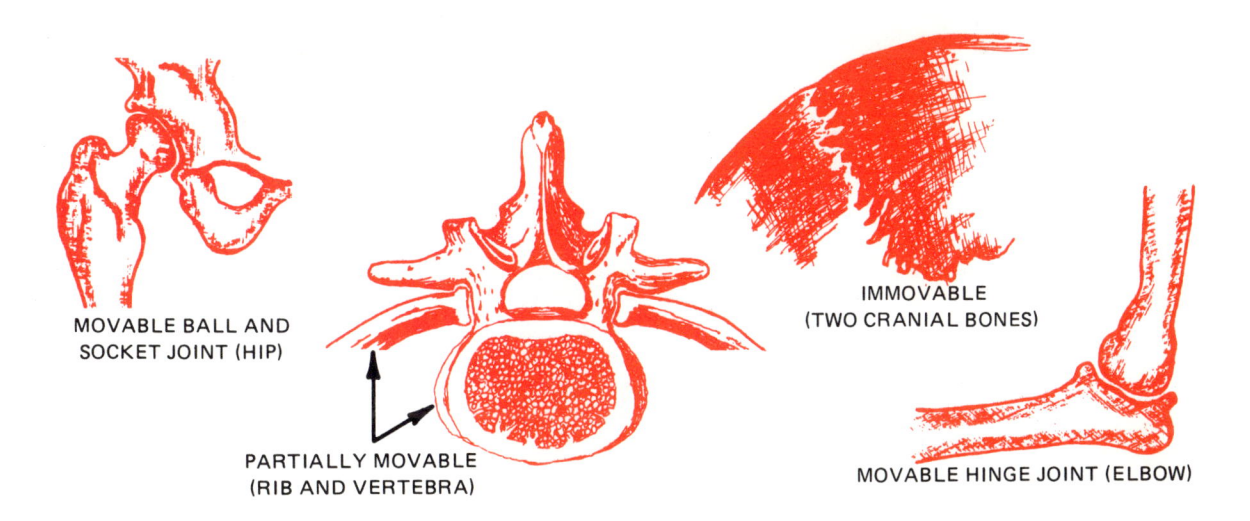

MOVABLE HINGE JOINT (ELBOW)

Fig. 5-2 Types of joints

true of the irregular bones of the spinal column. The thigh bone is a long bone, needed for support of the strong leg muscles and the weight of the body. The degree of movement at a joint is determined by bone shape and joint structure.

JOINTS AND RELATED STRUCTURES

Joints are points of contact between two bones. They may be movable, partially movable, or immovable, figure 5-2.

Movable joints may be one of several different types. *Ball and socket joints*, found in the shoulders and hips, give the greatest freedom of movement. *Hinge joints* move in one direction only, as in the knees, elbows, and two outer joints of the fingers. *Pivot joints* rotate one joint around the other joint as in the radius and ulna (long bones of the forearm). Another example is the joint between the atlas (first cervical vertebra which supports the head) and the axis (second cervical vertebra); this joint allows the head to turn. *Gliding joints* are those in which nearly flat surfaces glide across each other as between vertebrae of the spine. *Angular joints* (condyloid) provide movement in two directions as seen in wrist, ankle, and third finger-joints.

Partially movable joints (cartilaginous) are those in which bones are firmly jointed by cartilage as in the attachment of ribs to the spine. This allows for limited motion.

Immovable joints (fibrous) are found in the adult cranium. The bones are fused together in a joint which forms a heavy protective cover for the brain.

Ligaments are fibrous bands which connect bones and cartilages and serve as support for muscles. Joints are also bound together by ligaments. *Tendons* are fibrous cords which connect muscles to bones. Movable joints have a thin layer of cartilage covering the articulating surface. A smooth slippery membrane also covers movable joints and secretes a lubricating substance called *synovial fluid*. In the shoulder, knee and ankle, and knuckle joints there are tiny sacs of fluid called *bursa*. These tiny sacs act as a lubricating mechanism which cushions the joint.

TYPES OF MOTION

Joints can move in many directions, figure 5-3, page 16. *Flexion* is bending as

THE BODY FRAMEWORK

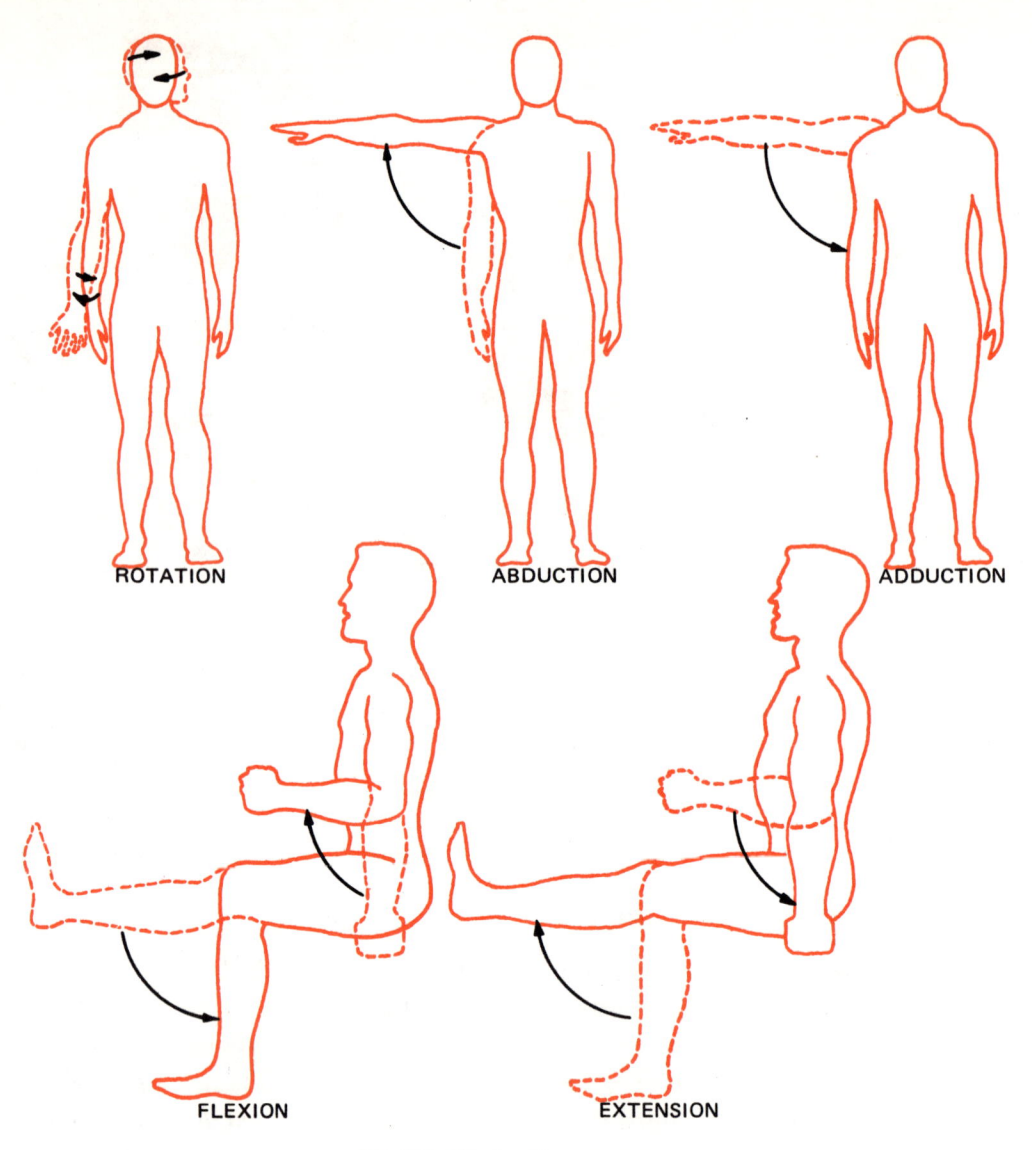

Fig. 5-3 Kinds of movements

when the forearm or fingers are bent or flexed. *Extension* means straightening the forearm or fingers. *Abduction* is the movement of an extremity away from the midline (imaginary line which halves the body from head to foot, dividing the left from the right side) while adduction is movement toward the midline. A *rotation* movement allows a bone to move around one central axis. Two rotation movements are *pronation,* in which the forearm turns so that the palm of the hand faces the body; and *supination,* in which the palm faces away.

SUGGESTED ACTIVITIES

- On a human skeleton model, point out the various joints which are movable and those which are immovable.

- Obtain the leg and wing of a turkey or chicken. Identify the bones, muscles, tendons, and ligaments. Note the toughness of the tendons

and ligaments. Observe the action of the joints, tendons, and ligaments as the leg is bent and straightened or the wing is spread.
- Discuss the differences in the structure of the male and female skeletons.
- Explain why tendons and ligaments are made of very tough tissue.

REVIEW

Select the letter of the answer which most correctly completes the statement.

1. Supination is one type of
    a. extension
    b. abduction
    c. adduction
    d. rotation

2. The bones found in the skull are
    a. irregular bones
    b. flat bones
    c. short bones
    d. long bones

3. The long bones are the site of
    a. storage of fat
    b. hormone secretions
    c. cartilage formation
    d. blood cell formation

4. The cranium protects the
    a. lungs
    b. brain
    c. heart
    d. stomach

5. Angular joints may be found in the
    a. vertebral column
    b. skull
    c. wrist
    d. shoulder

6. Irregular bones may be found in the
    a. leg
    b. vertebral column
    c. arm
    d. skull

7. Short bones are found in the
    a. leg
    b. arm
    c. vertebral column
    d. hand

8. Immovable joints are found in the
    a. infant's skull
    b. adult cranium
    c. adult spinal column
    d. child's spinal column

9. Flexion means
    a. bending
    b. rotating
    c. extending
    d. abduction

10. The degree of motion at a joint is determined by
    a. the amount of synovial fluid
    b. the number of bursa
    c. the unusual amount of exercise
    d. bone shape and joint structure

unit 6 structure and formation of bone

OBJECTIVES

Studying this unit should help the student:
- Explain the formation of bones.
- Describe bones with regard to composition and construction.
- Relate bone changes to body growth.

Bones are made up of microscopic cells which secrete large amounts of mineral matter; this makes bone tissue extremely hard. Calcium and phosphorus are important dietary elements which furnish cells with the necessary materials to manufacture mineral matter. Bones are 50 percent water and 50 percent solid materials. The inorganic substances in bones are: calcium phosphate, calcium carbonate, calcium fluoride, magnesium phosphate, sodium oxide, and sodium chloride.

FORMATION OF BONE

Bones begin to form in the early embryonic stage. Most bones replace previously formed cartilage. Some *ossification* (deposition of mineral matter and appearance of bone cells) begins soon after the second month in utero. Infant bones are very soft and pliable because of incomplete ossification at birth. Ossification due to mineral deposits continues through childhood. As bones ossify they become hard and more capable of bearing weight.

During infancy the bones are soft and are composed of hyaline cartilage which is covered by a fibrous membrane (thin lining or covering made of epithelial or connective tissue). The cartilage is gradually replaced by hard bone tissue as the child grows. A familiar example is the soft spot on a baby's head, called the *fontanel*. The bone has not yet been formed there but will become hardened later.

STRUCTURE OF BONE

A typical long bone is composed of a shaft, or *diaphysis*, and two extremities called *epiphysis*, figure 6-1. In the center of the shaft is the broad *medullary canal*. This is filled with yellow marrow, mostly fat cells. The marrow also contains many blood vessels and some cells which form white blood cells, *leukocytes*. The yellow marrow functions as a fat storage center. The marrow canal is lined and the cavity kept intact by the *endosteum*.

The medullary canal is surrounded by *compact* or hard bone. *Haversian canals* branch into the compact bone. They carry blood vessels which nourish the *osteocytes*, or bone cells. Where less strength is needed in the bone, some of the hard bone is dissolved away leaving *spongy* bone.

The ends of the long bones contain the red marrow where some red blood cells called *erythrocytes* and some white blood cells are made. The outside of the bone is covered with the *periosteum*, a tough fibrous tissue which contains blood vessels, lymph vessels, and nerves. The periosteum is necessary for bone growth, repair and nutrition.

GROWTH OF BONE

Bones grow in length at the epiphyseal extremities. The length of the shaft continues to grow until the epiphysis cells are hardened. Growth in females continues to about 18 years and in males to about 20 or 21 years.

STRUCTURE AND FORMATION OF BONE

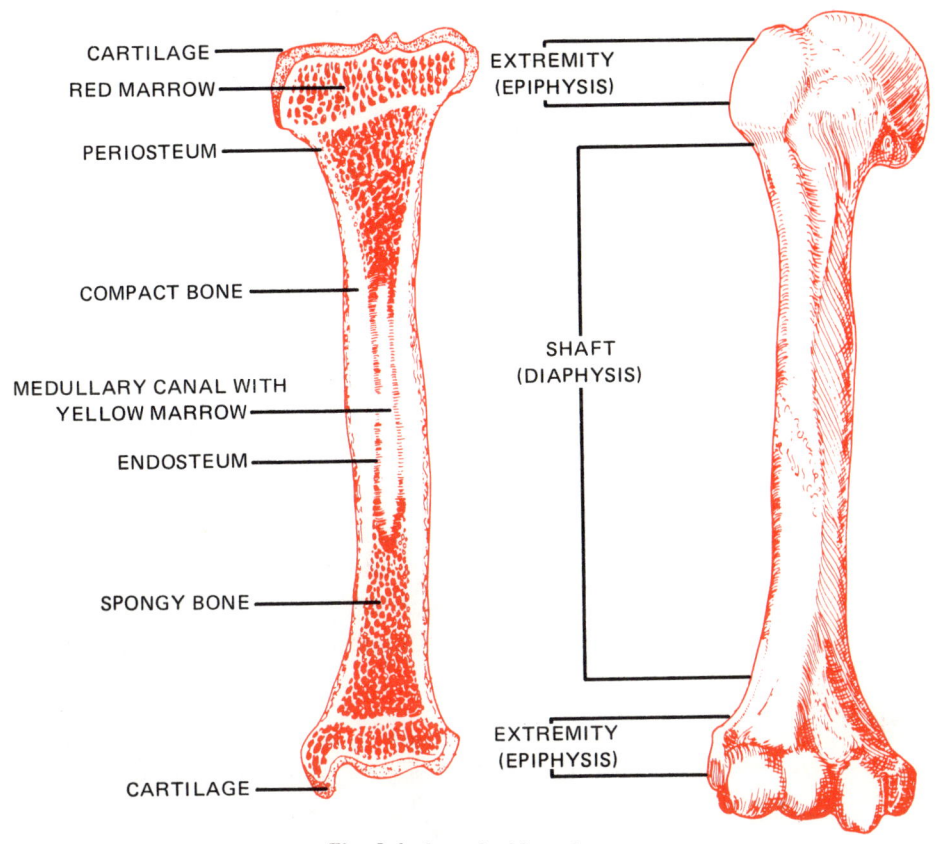

Fig. 6-1 A typical long bone

SUGGESTED ACTIVITIES

- Obtain a long beef bone from the butcher shop. Have it sawed through the center lengthwise so that the inner portion of the bone is visible. Identify each part of the bone.
- Discuss how the periosteum is involved in the growth of a bone.

REVIEW

A. Briefly answer the following questions.

1. Explain the difference between compact bone and spongy bone.

2. Explain the function of the red marrow of the bone.

THE BODY FRAMEWORK

3. Why do infant bones tend to be soft and pliable?

4. Discuss the pattern of bone growth.

B. Match each of the terms in column I with its correct statement in column II.

Column I	Column II
_____ 1. mineral matter	a. dietary elements which furnish cells with necessary materials to manufacture mineral matter
_____ 2. ossification	
_____ 3. fontanel	
_____ 4. endosteum	b. center of the bone shaft
_____ 5. calcium and phosphorus	c. part of bone containing yellow marrow, blood vessels, and some cells which form white blood cells
_____ 6. epiphysis	d. stage of development when bones begin to form
_____ 7. periosteum	
_____ 8. bone marrow	e. the process of mineral deposition and bone cell growth
_____ 9. medullary canal	
_____ 10. early embryonic period	f. sodium chloride
	g. area in infant skull where bone has not yet formed
	h. lining of the bone marrow canal
	i. elements which make bones hard and durable
	j. end structure of a long bone
	k. bone cells or osteocytes
	l. covering around bone which contains blood vessels, lymph vessels, and nerves

unit 7 parts of the skeleton

OBJECTIVES

Studying this unit should help the student:

- Name the components of the two main parts of the human skeleton.
- Describe the function of the main bone structures.
- Locate the bones in the human skeleton.

The skeletal system is composed of two main parts. The *axial* skeleton consists of the skull, spinal column, ribs, and breastbone. The second part, called the *appendicular* skeleton, consists of the arms, legs, shoulder girdle, and pelvic girdle, figure 7-1.

The skull, or head, consists of the *cranium* and the face. The flat bones which form the

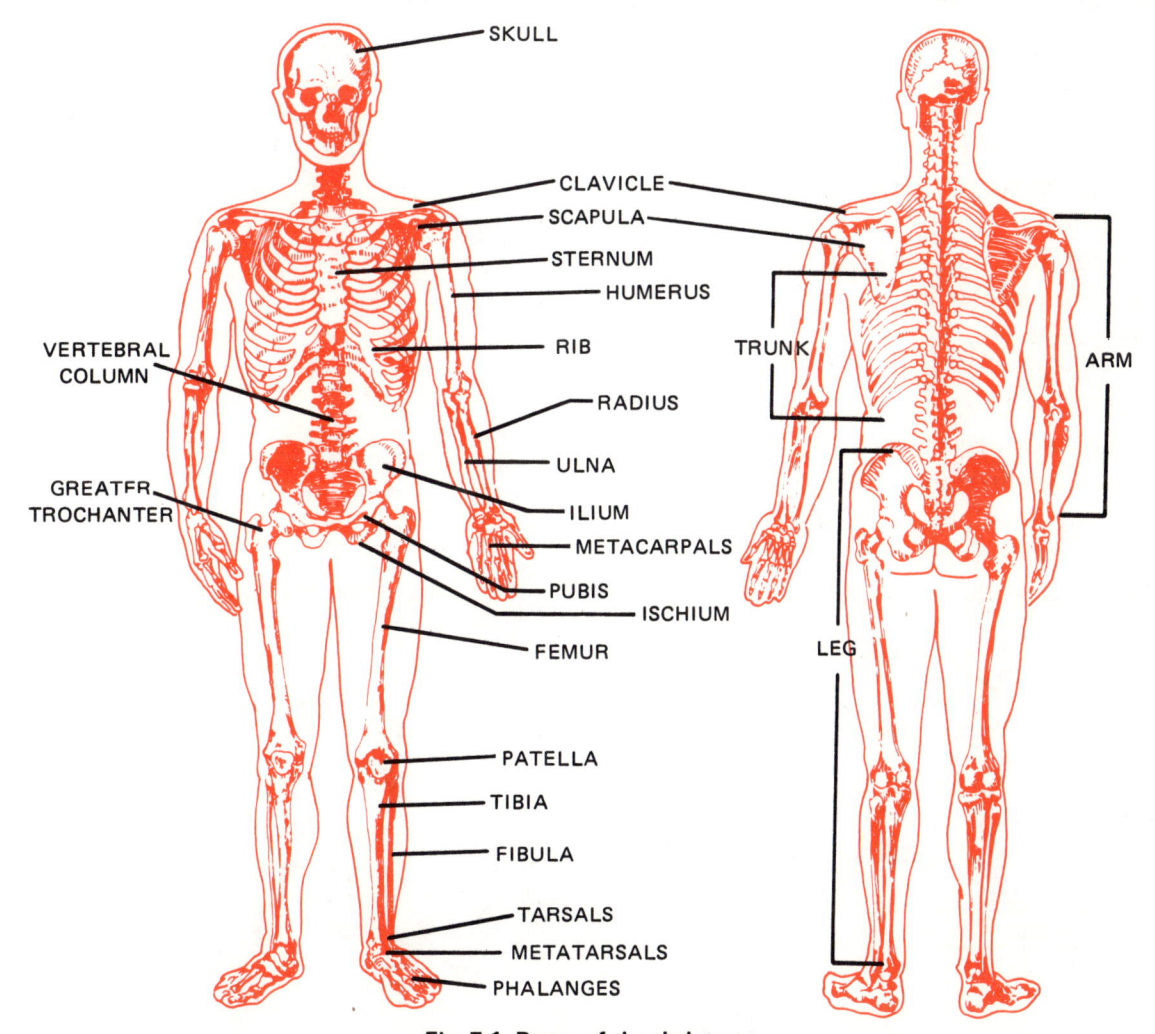

Fig. 7-1 Bones of the skeleton

21

The Body Framework

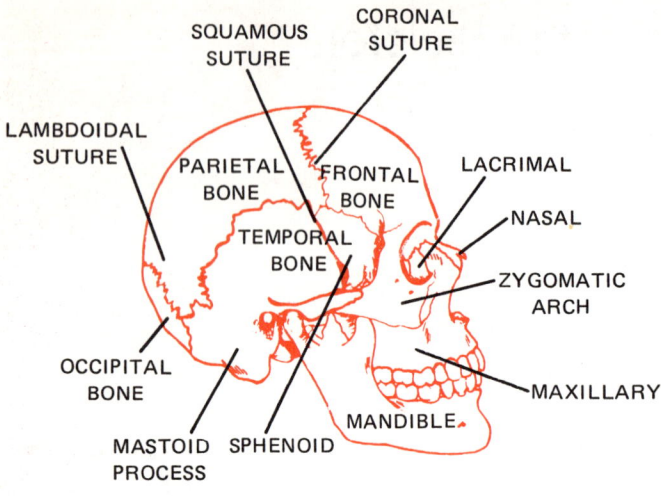

Fig. 7-2 The skull

cranium give protection to the delicate brain. Of the fourteen facial bones, the lower jaw (mandible) is the largest, figure 7-2. Three tiny bones of the ear, the hammer, anvil and stirrup, assist in the hearing function.

The spine, or *vertebral column,* is strong and flexible. It supports the head and provides for the attachment of the ribs. The spine also encloses the spinal cord of the nervous system.

The spine consists of small bones called *vertebrae* which are separated from each other by pads of cartilage tissue called intervertebral discs. The first vertebra is called the *atlas,* which supports the skull. The second is called the *axis.* It makes possible rotation of the skull on the neck as the atlas turns on the axis.

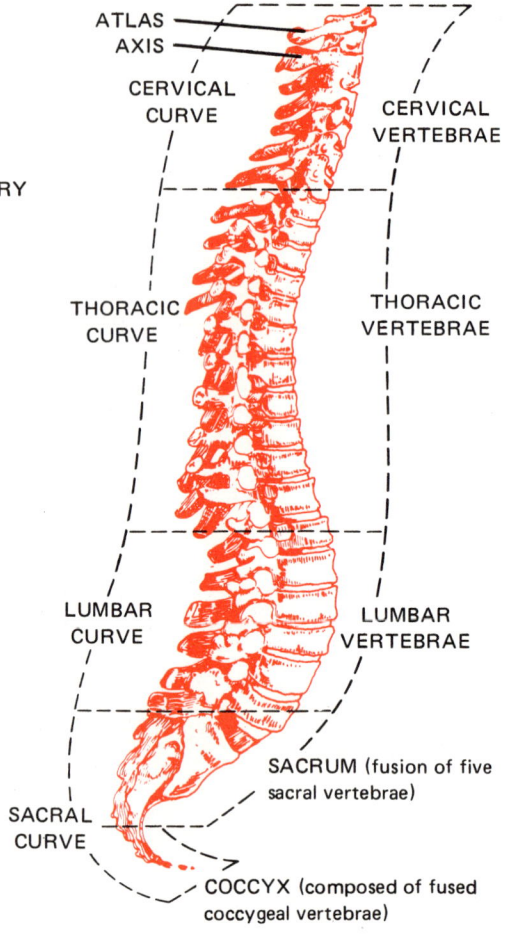

Fig. 7-3A The spinal column

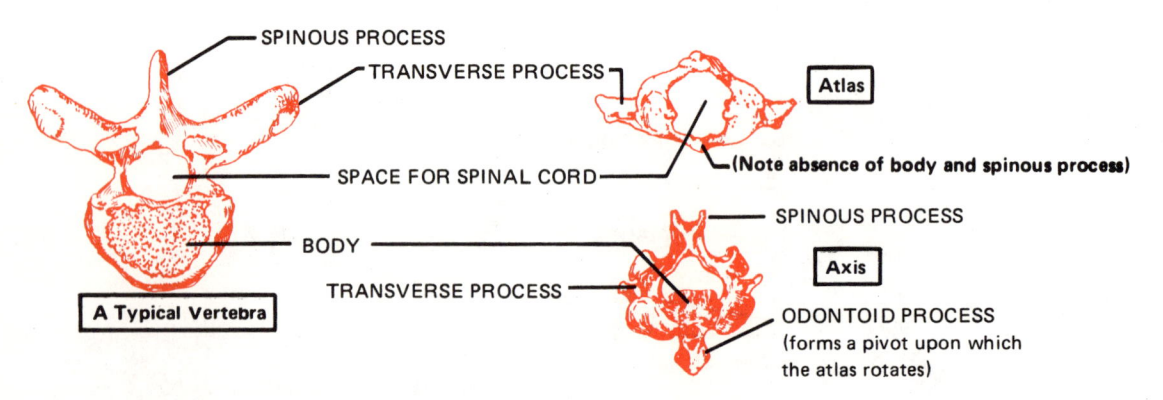

Fig. 7-3B Vertebrae comparison

PARTS OF THE SKELETON

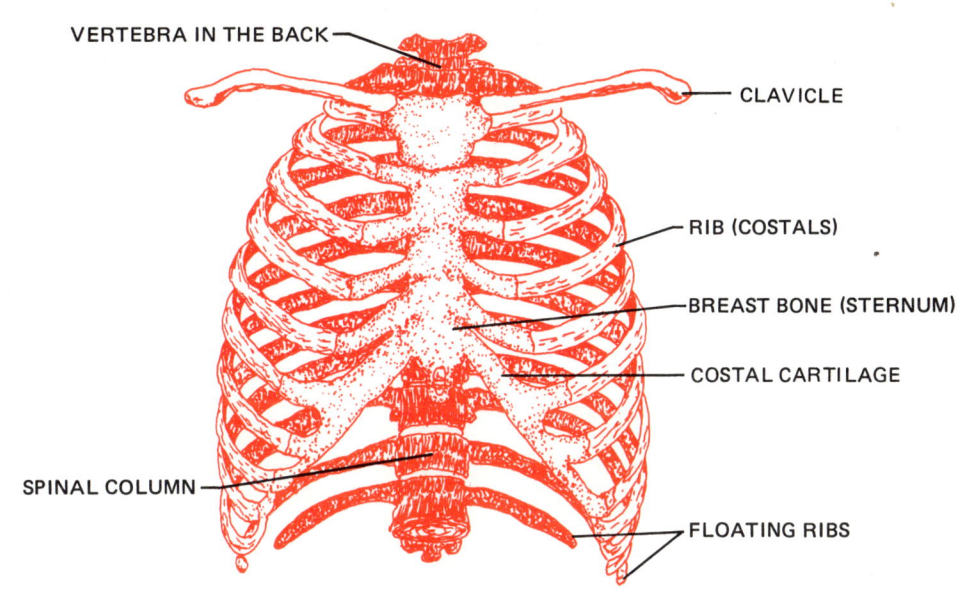

Fig. 7-4 Ribs and breast bone

Twelve pairs of ribs and a breastbone, or *sternum,* in the chest gives protection to the lungs and vital organs of the chest cavity.

The shoulder girdle consists of four bones (two *clavicles* and two scapulae) and serves as a place of attachment for the arms.

The *innominate* bones (ilium, ischium, pubis) form the pelvic girdle and provide attachment for bones and muscles of the legs. The pelvis is formed by the innominate bones, the *sacrum,* and the *coccyx.*

The bone structure of the arm consists of the *humerus* in the upper arm and the *radius* and *ulna* in the forearm.

The wrist or upper part of the hand has eight *carpal* bones. The hand has five *metacarpal* bones and the five fingers have fourteen bones called phalanges.

The upper leg contains the largest bone in the body, the *femur,* or thigh bone. The lower leg consists of the *tibia* and *fibula* bones. The ankle, or *tarsus,* contains seven *tarsal* bones; the foot, five *metatarsal* bones; and the five toes on the foot contain fourteen *phalanges,* figure 7-5.

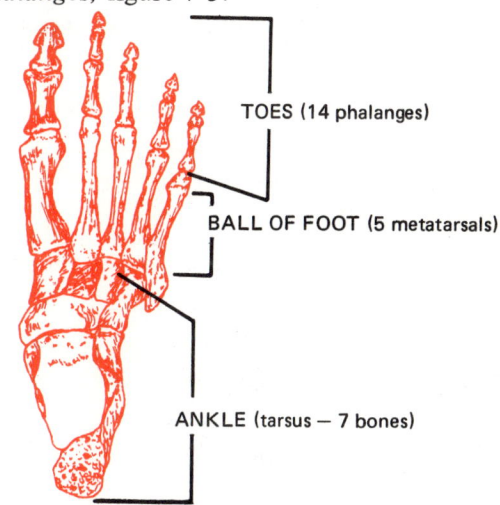

Fig. 7-5 Bones of the right foot

SUGGESTED ACTIVITIES

- Using a model of the human skeleton, identify the following: the cranium, face, shoulder girdle, pelvic girdle, spinal column, arms and legs; long bone, short bone, flat bone and irregular bone; the axis, the atlas, the cervical vertebra, the thoracic vertebra, the lumbar vertebra and the sacrum.

THE BODY FRAMEWORK

- Learn the meaning, spelling, and use of the following terms:

appendicular	frontal	nasal	temporal
axial	lacrimals	parietals	zygomatic
ethmoid	maxilla	sphenoid	

- Find out why the lowest two pairs of ribs are called the floating ribs.
- Explain why the first vertebra in the spinal column is called the atlas and why it has no body like other vertebrae.

REVIEW

Complete the following sentences.

1. The two main parts of the skeletal system are the _____ skeleton and the _____ skeleton.
2. Two main areas of the skull consist of the _____ and the _____ .
3. The largest of the fourteen facial bones is the _____ .
4. The three tiny bones of the ear which assist in the hearing function are the _____ , _____ , and _____ .
5. The spinal or vertebral column consists of small bones separated from each other by pads of _____ called _____ _____ .
6. The odontoid process is an important structure of the second vertebra, the _____ .
7. The flat bone lying between the ribs in the front of the chest is the _____ .
8. The bones which form the pelvic girdle are called the _____ bones.
9. The individual bones which form the pelvis are the _____ , _____ , _____ , _____ , and _____ .
10. The largest bone in the body is the _____ or _____ bone.
11. The bones of the skull are the _____ , _____ , _____ , and _____ bones.
12. The sutures of the skull are the _____ , _____ , and _____ sutures.
13. The spinal column has five main vertebral sections, the _____ , _____ , _____ , _____ , and _____ .
14. The first vertebra which supports the skull is the _____ .
15. The odontoid process forms a pivot upon which the _____ vertebra rotates.

unit 8 injuries and diseases of bones and joints

OBJECTIVES

Studying this unit should help the student:

- Define four types of bone fractures.
- Identify common bone and joint injuries.
- Identify common bone and joint disorders.

The most common injury to a bone is a *fracture,* or break. When this occurs, there is swelling due to injury and bleeding tissues. The process of restoring the fractured bone to its original position is known as *reduction*. A cast is put on to hold the fracture in place and at rest. Healing takes place and the bones knit, or grow, together again. The following chart identifies the common types of fractures.

SIMPLE	Bone is broken but there is no external wound. Skin has not been pierced.
COMPOUND	Bone is broken; the external wound leads to site of the fracture. Skin has been broken and bone may have pierced through.
GREENSTICK	Bone is partly bent and partly broken as when a green stick breaks. This is especially common in children's fractures.
COMMINUTED	Bone is splintered and pieces of bone become imbedded in surrounding tissue.

A *dislocation* occurs when a bone is displaced from its proper position in a joint. This may result in the tearing and stretching of the ligaments. Reduction or return of the bone into position is necessary along with rest to allow the ligaments to heal.

A *sprain* is an injury to a joint caused by any unusual strain, such as "turning the ankle." The ligaments are either torn from their attachments to the bones or torn across, but the joint is not dislocated. Treatment consists of supporting the joint until the ligaments heal. This is usually done with adhesive strapping.

Arthritis is an inflammatory condition of one or more joints. It is usually accompanied by pain and often by changes in bone position.

A *bunion* is the swelling of the bursa of the foot, usually in the big toe. It results from poorly fitting shoes, poor walking posture or genetic tendency.

Rickets is a disease of the bones which is caused by lack of vitamin D. Portions of the bones are soft, due to lack of calcification. The soft bones bend, causing such deformities as bowlegs and pigeon breast. The disease may be prevented by providing a growing child with sufficient quantities of calcium, vitamin D, and exposure to sunshine.

Clubfoot (talipes) is a congenital (existing at birth) malformation. It may involve one or both feet. The deformity may take one of several forms; the body weight may rest on the heel or ball of the foot only, or with only the inner or outer side of the sole touching the ground.

Microcephalus is a congenital or hereditary (inborn) condition in which there is a marked diminution in the size of the cranium due to early ossification of the sutures of the skull. This is usually accompanied by arrested mental development.

Spina bifida is a congenital condition in which the vertebral canal, which contains the spinal cord, did not develop completely and unite properly. This condition usually occurs in the lumbar and sacral regions. Frequently the contents of the spinal canal protrude.

Osteoporosis, or softening of the bones, is a condition caused by a deficiency in male or female hormones. The bones fracture easily because of brittleness. Replacement of hormones and increased mineral intake may slow the process of mineral withdrawal.

Bursitis is inflammation of the bursae which cushion the joints during motion. Early medical treatment may save the patient weeks of pain as well as inability to use the joint.

Webbed and extra fingers and toes are congenital conditions which must be corrected by early surgery.

Scoliosis is a side-to-side or lateral curvature of the spine.

Kyphosis is a humped curvature in the thoracic area of the spine.

Lordosis is an exaggerated inward curvature in the lumbar region of the spine just above the sacrum.

Bone disorders also occur with *tuberculosis, osteomyelitis,* and benign or malignant *tumors.*

SUGGESTED ACTIVITIES

- Make drawings of the different types of fractures as they would look if the leg were broken.

REVIEW

Briefly answer the following questions.

1. What is the difference between a simple and a compound fracture?

2. Explain why a child with rickets would have bowlegs and other bone deformities. How can this disease be prevented?

section 2 the body framework
SELF-EVALUATION

A. Match each of the terms in column I with its correct statement in column II.

Column I	Column II
_____ 1. atlas	a. also called innominate bones
_____ 2. axis	b. bone of the upper arm
_____ 3. femur	c. fibrous band which joins bone to bone
_____ 4. frontal bone	d. contents of medullary canal
_____ 5. humerus	e. necessary for growth and repair of bone tissue
_____ 6. ligament	f. name for thigh bone
_____ 7. mandible	g. vertebra on which skull rests
_____ 8. pelvic girdle	h. name for forehead
_____ 9. periosteum	i. name of lower jaw
_____ 10. red marrow of bone	j. vertebra on which head rotates
_____ 11. tendon	k. makes red blood cells
_____ 12. yellow marrow of bone	l. joins a muscle to a bone
	m. name of upper jaw

B. State five functions of the skeletal system.

C. State the difference between a dislocation and a sprain.

D. Explain the location and functions of the periosteum and Haversian canals.

27

SECTION 3

BODY MOVEMENT

unit 9 introduction to the muscular system

OBJECTIVES

Studying this unit should help the student:

- Describe the functions of muscles.
- Describe each of the three muscle types.

The body is supported by bones that are moved by muscles pulling on them; the joints facilitate bending. Muscles serve important functions in addition to moving the body or any of its parts. They help keep the body erect and determine posture. They also produce most of the heat that is generated in the body.

Muscles are made of bundles of muscle fibers held together in varying numbers and lengths by connective tissue. Muscle tissue has the ability to contract and to shorten or become tense and then return to its original size.

There are three principal types of muscles. *Voluntary muscles* are made of long striped cells, figure 9-1. They attach to the bony skeleton and move its parts. They can be controlled by the will. *Involuntary* or *smooth muscles* are composed of small spindle-shaped cells. They are found in the walls of the internal organs, principally the stomach, intestines, and walls of blood vessels. Involuntary muscles are generally not controlled by the will but by the automatic activities of the nervous system. *Cardiac muscle* is made of special branching muscle cells found only in the heart, figure 9-3.

Sphincter muscles are special circular muscles such as the anus and the urethra which open and close to control the passage of substances. These sometimes work either voluntarily or involuntarily.

Fig. 9-1 Voluntary or striated muscle cells

Fig. 9-2 Involuntary or smooth muscle cells

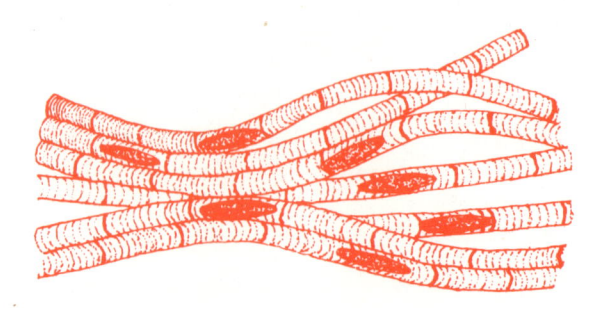

Fig. 9-3 Cardiac muscle cells

INTRODUCTION TO THE MUSCULAR SYSTEM

SUGGESTED ACTIVITIES

- Observe prepared slides of muscle tissue or bring several types of uncooked meat to class (tripe, steak and heart). Using dissecting needles, place a few fibers on a slide in a drop of water. Observe each under the compound microscope. Notice the different kinds of fibers. Draw and label these fibers. Describe how they differ in appearance.
- Make a chart listing the 3 muscle types and several examples of each type.

REVIEW

Place the most correct answer in the space or spaces provided.

1. Muscles help to keep the body erect and therefore determine our _____ .

2. Muscles produce most of the _____ that is generated in the body.

3. The action of muscles upon bones is responsible for movements of our _____ .

4. A specialized muscle, the heart, is responsible for _____ throughout the body.

5. Muscle tissue helps get carbon dioxide out and oxygen into the body through the _____ .

6. Three principal types of muscle tissue are the _____ , _____ , and the _____ .

7. Cardiac muscle is an _____ muscle.

8. Circular muscles found at the anus and urethra are called _____ muscles; they control the _____ of excretion from these areas.

9. Muscular tissue is composed of bundles of _____ _____ held together by _____ _____ .

10. Muscular tissue has the ability to _____ , _____ and _____ as needed and the ability to _____ to its original size.

29

unit 10 attachment of muscles

OBJECTIVES

Studying this unit should help the student:

- Describe how pairs of muscles work together.
- Describe how muscles are attached.
- Describe how muscles are ready for action.

There are over 500 different muscles in the body. They are arranged in pairs, one muscle being antagonistic to the other and performing an action opposite to that of the other. The muscle which bends a joint is called a *flexor*. It appears thicker and shorter than the muscle which straightens the joint, the *extensor*. There is actually no change in the amount of muscle tissue during movement but merely a shortening and "bunching up" of the fibers, figure 10-1.

Muscles are attached at both ends to bones, cartilage, ligaments, tendons, skin, and sometimes to each other. The end which moves least during muscle contraction is the *origin;* the end moving most is the *insertion*. The *belly* is the body of the muscle.

The blood circulation, body tissues, and the liver supply oxygen and sugar from digested food necessary for the work of the muscle cells. Muscles and the liver store sugar as glycogen which they can convert to glucose.

Muscles give off heat and waste following exercise. Muscle *fatigue* is caused by an accumulation of lactic acid which is a result of incomplete oxidation of the stored sugar. *Oxidation,* which means combining with oxygen, is the main chemical process by which energy is released from sugar metabolism in the mitochondria of the cells.

In order to function rapidly and well, muscles should always be slightly contracted and ready to pull, even when they are not pulling. This is *muscle tone*. People in good health have firm muscles which are always ready to work. Muscle tone can be achieved through proper nutrition and regular exercise.

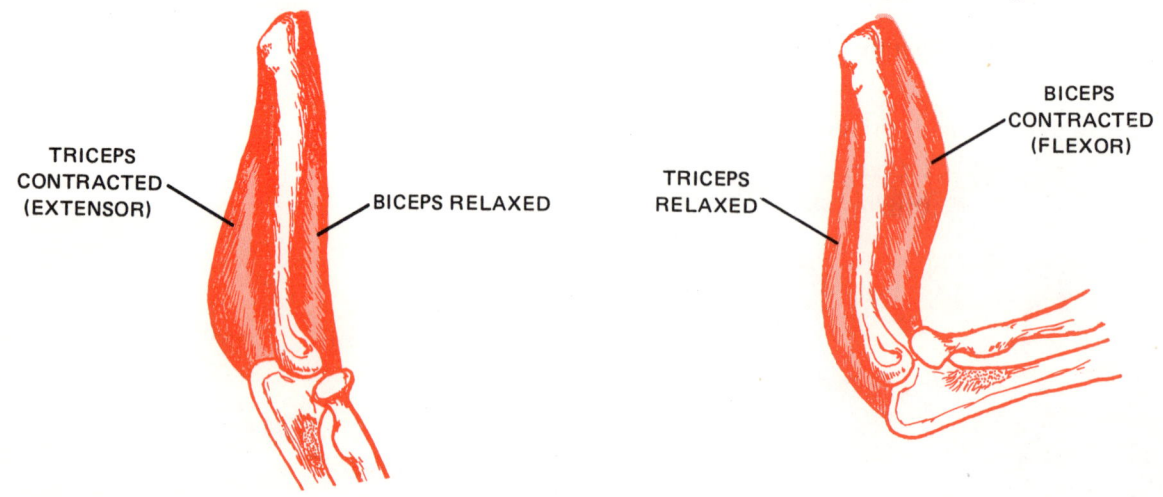

Fig. 10-1 Coordination of antagonistic muscles

ATTACHMENT OF MUSCLES

Each muscle is in contact with the nervous system through the motor nerve which carries messages from the brain to the muscles and makes them always ready for action.

SUGGESTED ACTIVITIES

- Ask your butcher for a chicken leg with the foot still intact. Cut open the skin and free the ends of the tendons. Pull the tendons in the front of the leg, then those behind it. What happened to the toes? Did the tendons stretch?
- Discuss what happens to muscles that are not used.
- Make arrangements for a physiotherapist to speak about maintaining muscle tone in the elderly patient.

REVIEW

Complete the following sentences.

1. Muscles are arranged in _____ , one muscle being _____ to the other and performing an action _____ to that of the other.
2. A muscle which bends a joint is called a _____ .
3. The muscle which bends a joint appears thicker and shorter than the one that straightens the joint, called an _____ .
4. Muscles may be attached to _____ , _____ , _____ , _____ , _____ , and sometimes to _____ _____ .
5. The end of the muscle which moves least during muscle contraction is the _____ .
6. The end of the muscle moving the most is the _____ .
7. The nourishment necessary for the work of the muscle cells is furnished by the _____ and carried by the _____ .
8. Muscles and the liver store _____ which may be converted to glucose.
9. Muscle fatigue is caused by an accumulation of _____ _____ , a result of _____ _____ of stored sugar.
10. Oxidation is the main _____ _____ by which _____ is released from the sugar in the cells.

unit 11 principal skeletal muscles

OBJECTIVES

Studying this unit should help the student:

- Locate the important skeletal body muscles.
- Describe the function of these muscles.
- Identify the muscles using the technical names.

The diagram in figure 11-1 lists the principal skeletal muscles, their locations and functions. The muscle groups as they appear on the human body can be studied in figure 11-2.

MUSCLE	LOCATION	FUNCTION
Sternocleidomastoid	Neck	Moves head
Deltoid	Shoulder	Abducts upper arm
Biceps (brachii)	Upper arm	Flexes lower arm
Triceps (brachii)	Upper arm	Extends lower arm
Pectoralis major	Anterior chest	Flexes upper arm; Helps adduct upper arm
Intercostals	Between ribs	Move ribs (assist in breathing)
Diaphragm	Between abdominal and chest cavities	Enlarges thorax (assists in breathing)
Serratus	Anterior chest	Moves shoulder
Rectus abdominus	Extends from ribs to pelvis	Compresses abdomen
Sartorius	Anterior thigh	Flexes and rotates thigh and leg
Rectus femoris	Anterior thigh	Flexes thigh; Extends lower leg
Vastus lateralis	Anterior thigh	Extends leg
Tibialis anterior	Anterior leg	Flexes and elevates foot
Extensors	Ankle and foot	Move foot and toes
Extensors Flexors	Forearm	Move hand and fingers
Trapezius	Posterior chest and back	Moves shoulder; Extends head
Latissimus dorsi	Posterior chest and back	Extends upper arm; Helps adduct and rotate upper arm
Gluteus medius	Buttocks	Abducts and rotates thigh
Gluteus maximus	Buttocks	Extends thigh and rotates it outward
Biceps femoris	Posterior thigh	Flexes leg; Extends thigh
Gastrocnemius	Posterior leg	Points toes; Flexes lower leg
Sacrospinalis	Inserted in ribs and vertebrae	Extends spine; Abducts and rotates trunk

Fig. 11-1 Location and function of principal skeletal muscles

PRINCIPAL SKELETAL MUSCLES

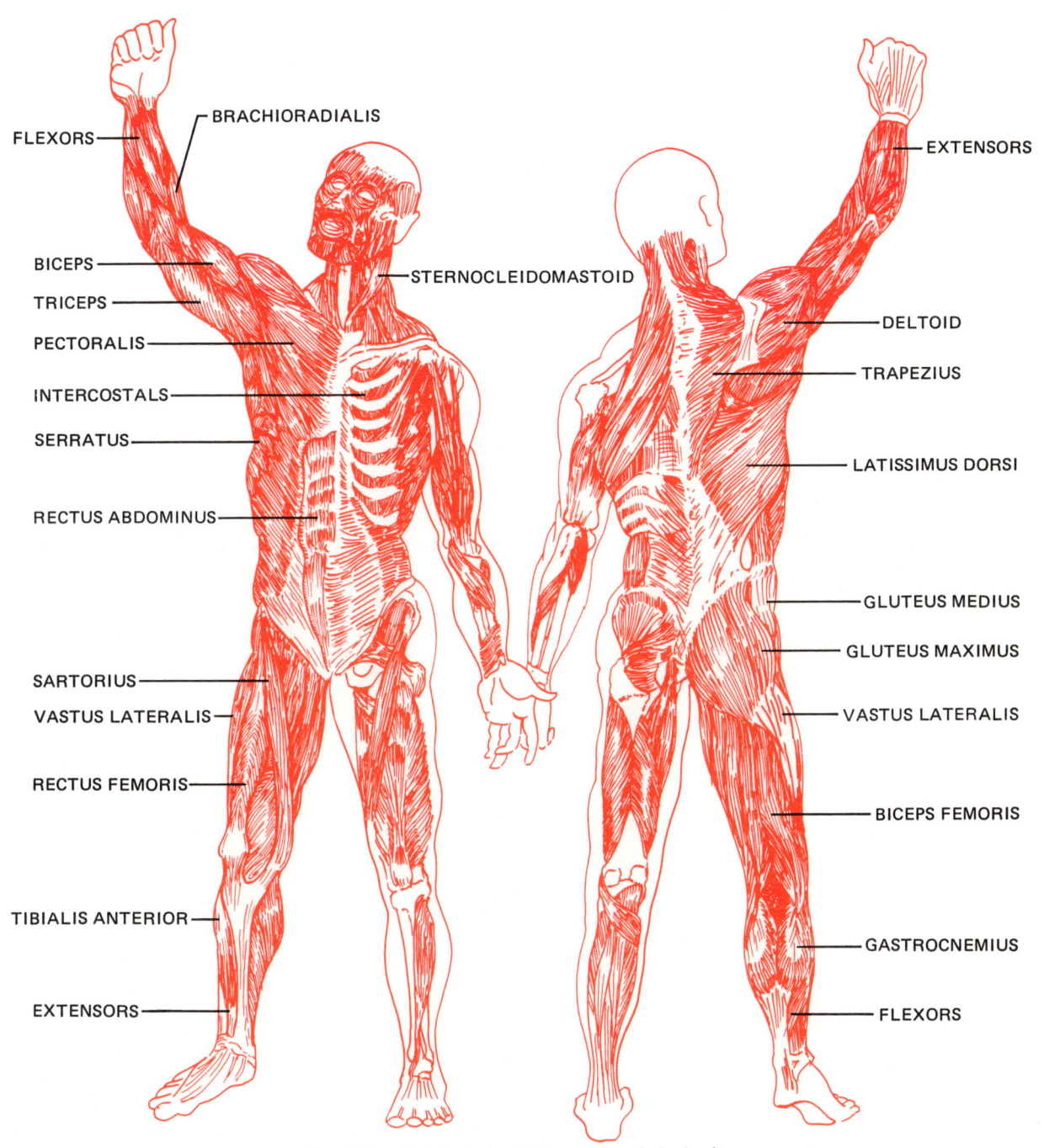

Fig. 11-2 Principal skeletal muscles of the body

SUGGESTED ACTIVITIES

- Discuss which muscles are used for giving injections.
- Identify which muscles seem to work in pairs.
- Discuss the effect of massage on muscles.

BODY MOVEMENT

REVIEW

Give the general function of the muscles named below.

1. Deltoid

2. Intercostals

3. Rectus Femoris

4. Gluteus Maximus

5. Triceps (brachii)

6. Pectoralis major

7. Sartorius

8. Extensors

9. Trapezius

10. Tibialis anterior

unit 12 abnormal muscular conditions

OBJECTIVES

Studying this unit should help the student:

- Identify some common abnormal conditions and diseases of the muscles.
- Describe how conditions and diseases may prevent proper muscle function.

Muscular coordination is very important if a person is to perform his daily functions efficiently. Injuries and diseases which may affect muscles sometimes interfere with these functions. The retraining of injured or unused muscles is a type of *rehabilitation,* called therapeutic exercise.

Flatfoot is a condition in which the muscles that support the arch are not able to meet the strain which is placed upon them. The strength of the muscles can be increased by exercise, massage, and electrical stimulation.

An *abdominal hernia,* or rupture, may occur in a weak place in the muscular abdominal wall. It is caused by bulging of the intestine through an opening in the wall of the abdominal cavity normally containing it. The *inguinal hernia* is the most frequent type of hernia. It appears in the groin area.

Muscle atrophy is a major loss of muscle strength and size. This may be caused by paralysis of nerves to the area or by lack of use of the muscle.

Muscle hypertrophy is a condition in which the muscle enlarges due to overwork. The heart muscle hypertrophy occurs after long-term overworking.

Muscle fatigue may occur from the temporary overuse of a muscle. Fatigue lessens the muscle's ability to perform work.

Paralysis is a condition which occurs when muscles do not receive nerve messages. Immobility often results.

Stiff neck is often caused from an inflammation of the trapezius muscle. The rigidity results from an unusual overuse of the muscle.

Tetanus, or lockjaw, is a continuous spasm of the muscles due to the toxins from the tetanus bacillus.

Muscular dystrophy is a chronic wasting disease of the muscles. It often occurs during childhood and is thought to result from some genetic disturbance.

Muscle spasms are sudden and violent contractions caused by sudden overworking of the muscle or by poor circulation to the area.

SUGGESTED ACTIVITIES

- Visit a physical rehabilitation center. Report on the success of muscle retraining in this program.
- Discuss how neglectful hospital care can lead to muscular atrophy in a patient.

REVIEW

Match each of the terms in column I with its correct statement in column II.

Column I	Column II
_____ 1. muscular atrophy	a. the retraining or rehabilitation of muscle use
_____ 2. muscular dystrophy	b. the temporary overuse of a muscle
_____ 3. paralysis	c. chronic wasting of the muscle tissue
_____ 4. stiff neck	d. sudden and violent muscle contraction
_____ 5. muscle fatigue	e. continuous spasm caused by a toxin
_____ 6. muscle hypertrophy	f. immobility caused by blocked nerve messages
_____ 7. muscle spasm	g. bulging of an organ through a muscular wall
_____ 8. hernia	h. major loss of muscle strength and size
_____ 9. therapeutic exercise	i. rigidity often caused from inflammation of the trapezius muscle
_____ 10. tetanus	j. muscle enlargement due to overworking of the muscle
	k. condition resulting from weak arch muscles

section 3 body movement
SELF-EVALUATION

A. Match each of the terms in column I with its correct statement in column II.

Column I	Column II
_____ 1. biceps	a. extends upper arm
_____ 2. diaphragm	b. assists in breathing
_____ 3. gastrocnemius	c. flexes lower arm
_____ 4. gluteus medius	d. extends spinal column and moves trunk
_____ 5. latissimus dorsi	e. moves foot and leg
_____ 6. sacrospinalis	f. closes body openings
_____ 7. serratus	g. abducts and rotates thigh
_____ 8. sphincter	h. moves shoulder
	i. moves the head

B. Define the following terms.

1. Flatfoot

2. Bursitis

3. Hernia, or rupture

4. Muscular atrophy

5. Paralysis

6. Tetanus

SECTION 4

TRANSPORT OF FOOD AND OXYGEN

unit 13 introduction to the circulatory system

OBJECTIVES

Studying this unit should help the student:
- Describe the function of the circulatory system.
- List the components of the circulatory system.
- Describe the two routes of blood circulation.

The circulatory system is the transportation system of the body. The bloodstream carries needed food, oxygen, and other substances to the tissues. It also carries waste products to the various organs of excretion where they are expelled from the body.

The blood circulatory system consists of the heart, arteries, veins, and capillaries. It also includes the lymphatic system which consists of the lymph and tissue fluid derived from the blood and the lymphatic vessels which return the lymph to the blood. The spleen is considered a part of the circulatory system. It provides a reservoir for blood and is active in destroying microorganisms in the blood. The spleen also forms some white blood cells, and removes old red blood cells from the blood.

The blood leaves the heart through arteries and returns by veins. The blood uses two routes of circulation:
1. The *general,* or *systemic, circulation* carries blood throughout the body, figure 13-1.
2. The *pulmonary circulation* carries blood from the heart to the lungs and back, figure 13-2. The blood circulates throughout the body very rapidly. Completion of both circuits of the blood takes only about one minute.

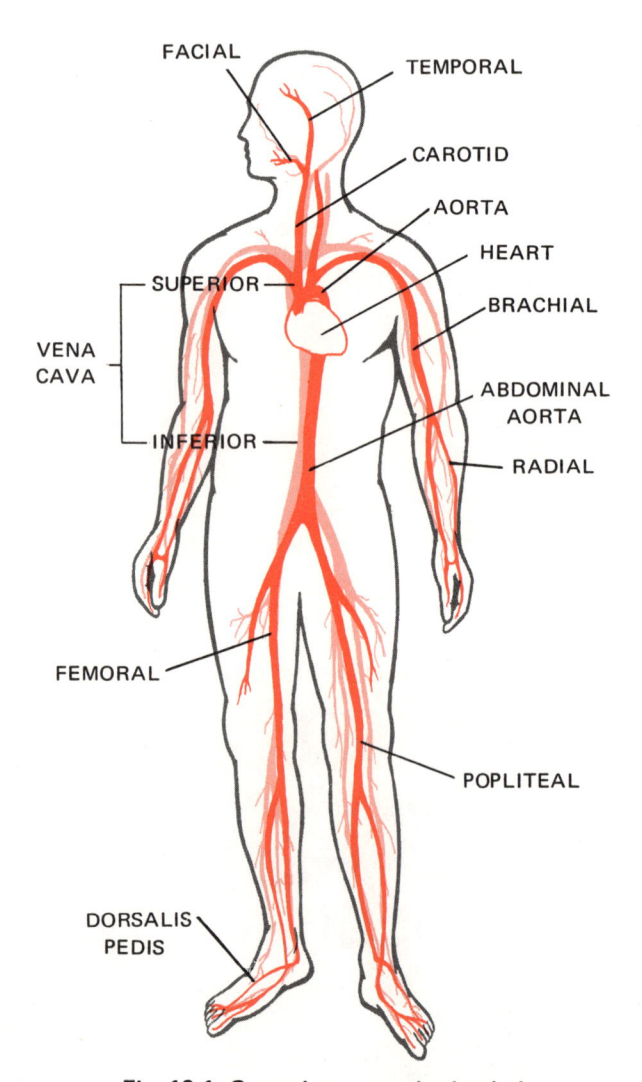

Fig. 13-1 General or systemic circulation

INTRODUCTION TO THE CIRCULATORY SYSTEM

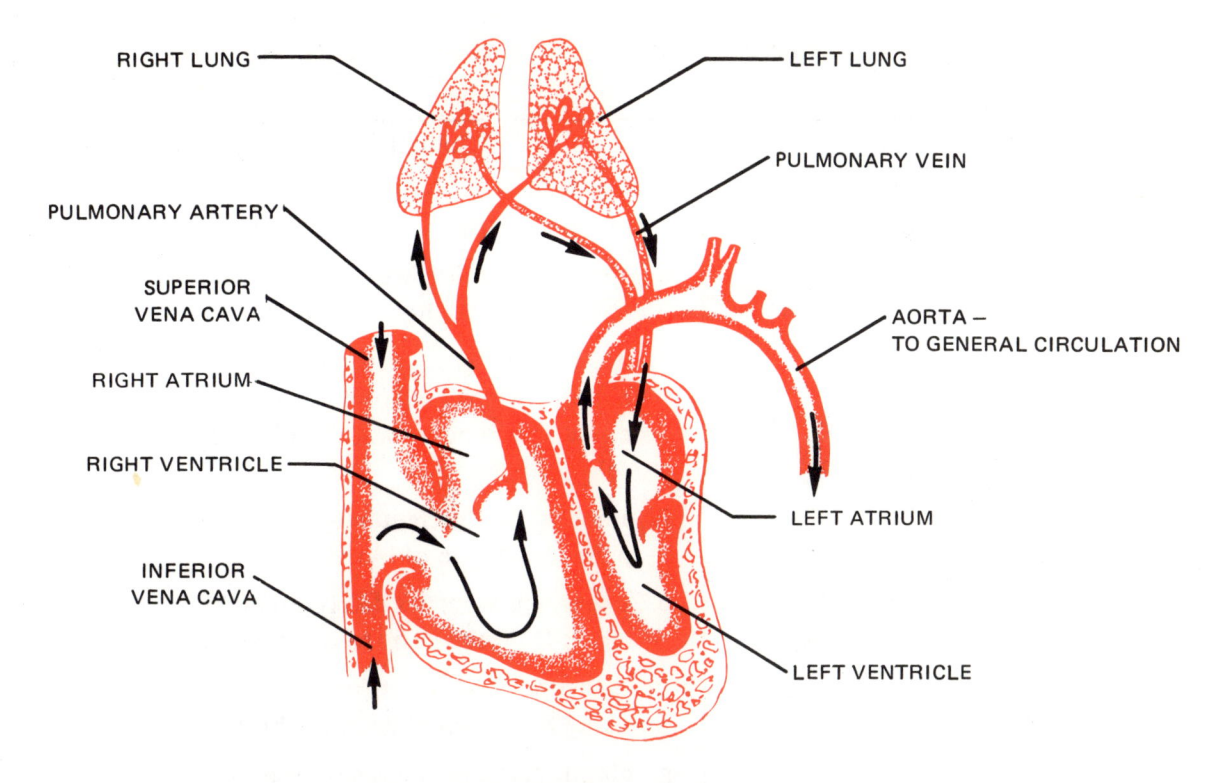

Fig. 13-2 Pulmonary circulation

SUGGESTED ACTIVITIES

- If available, observe blood cells circulating through capillaries in the web of a frog's foot. Note the direction of the circulation flow.
- Discuss how blood from the general circulation is oxygenated by blood from the pulmonary circulation.

REVIEW

A. Answer the following questions briefly.

1. Describe the chief functions of the circulatory system.

2. What is the name of the large vein which returns blood to the heart from the general circulation?

TRANSPORT OF FOOD AND OXYGEN

B. Match each of the terms in column I with its correct statement in column II.

Column I	Column II
_____ 1. pulmonary artery	a. vein which carries freshly oxygenated blood from the lung to the heart
_____ 2. lymphatic system	b. circulation route which carries blood to and from the heart and lungs
_____ 3. about one minute	c. organ which provides a reservoir for blood, destroys microorganisms in the blood and removes worn out blood cells
_____ 4. pulmonary vein	d. amount of time required for the complete circulation of blood
_____ 5. spleen	e. artery which carries deoxygenated blood from the heart to the lung
_____ 6. pulmonary circulation	f. system which consists of lymph and tissue fluid derived from the blood
_____ 7. left ventricle	g. blood from the pulmonary vein re-enters the heart through the right atrium
_____ 8. general circulation	h. artery which carries blood with nourishment, oxygen and other materials from the heart to all parts of the body
_____ 9. right ventricle	i. ventricle from which the aorta receives blood
_____ 10. aorta	j. organ which destroys microorganisms in the blood and removes worn out cells
	k. circulation which carries blood throughout the body
	l. ventricle from which the pulmonary artery leaves the heart

unit 14 the heart

OBJECTIVES

Studying this unit should help the student:

- Describe the structure of the heart.
- Describe the function of various parts of the heart.
- Locate and identify various parts of the heart.

The blood circulatory system, like other systems of the body, is extremely efficient. The main organ responsible for this efficiency is the heart, a tough, simply-constructed muscle about the size of a closed fist.

The heart is a four-chambered, muscular pump which keeps the blood circulating through the blood vessels to all parts of the body. It is conical in shape and is located between the lungs in the center and lower part of the chest cavity. It is enclosed in a sac called the *pericardium*, a protective covering which lessens the friction caused by the heartbeat. The smooth lining of the heart is called the *endocardium*. This tissue provides smooth transit for the flowing blood and also forms the valves.

The four chambers of the heart are the right atrium, the right ventricle, the left atrium, and the left ventricle. The *atrium* is sometimes called *auricle*. The heart has four valves which permit the blood to flow in one direction only. These valves are located as follows and prevent the blood from flowing backward into the chambers, figure 14-1, page 42.

1. The *mitral*, or *bicuspid*, is between the left atrium and the left ventricle.
2. The *tricuspid* is between the right atrium and the right ventricle.
3. The *pulmonary semilunar* is at the orifice of the pulmonary artery.
4. The *aortic semilunar* is at the orifice of the aorta.

The heart muscle is called the *myocardium* or the cardiac muscle. The myocardium contracts rhythmically in order to perform its duty as a forceful pump. The control of the contractions of the heart muscle is found in the *sinoatrial* (S-A) node located at the opening of the superior vena cava into the right atrium, figure 14-2, page 42. The S-A node is called the *pacemaker* of the heart. It cannot be identified except by microscopic examination of the cardiac tissue. A device called the electric cardiac pacemaker has successfully maintained a normal heartbeat under conditions of *cardiac arrhythmia* (irregular heart action).

Contractions of the atriums stimulate a second node into activity. This is the *atrioventricular* (A-V) node which is located where the atriums and ventricles meet. A side branch from the A-V node enters each ventricle through the *bundles of His*. Tiny fibers from these branches form a network immediately below the endocardium. Thus an impulse passes through these fibers of the ventricles until it reaches the apex of the heart.

The combined action of the S-A and A-V nodes is instrumental in the cardiac cycle. The cardiac cycle comprises one complete heartbeat, with both atrial and ventricular contractions.

1. The S-A node stimulates the contraction of both atriums, and blood flows from the atriums into the ventricles through the

TRANSPORT OF FOOD AND OXYGEN

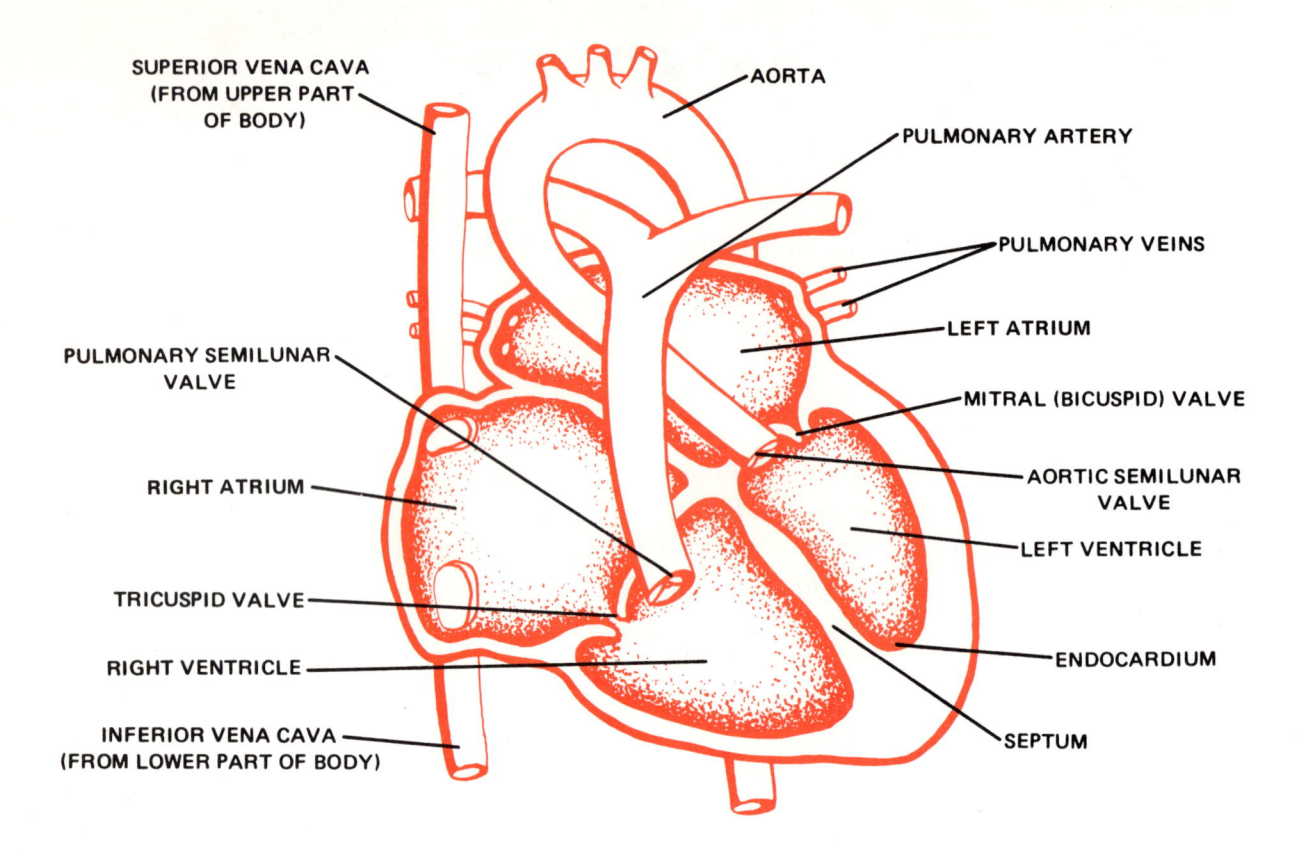

Fig. 14-1 The heart and its valves

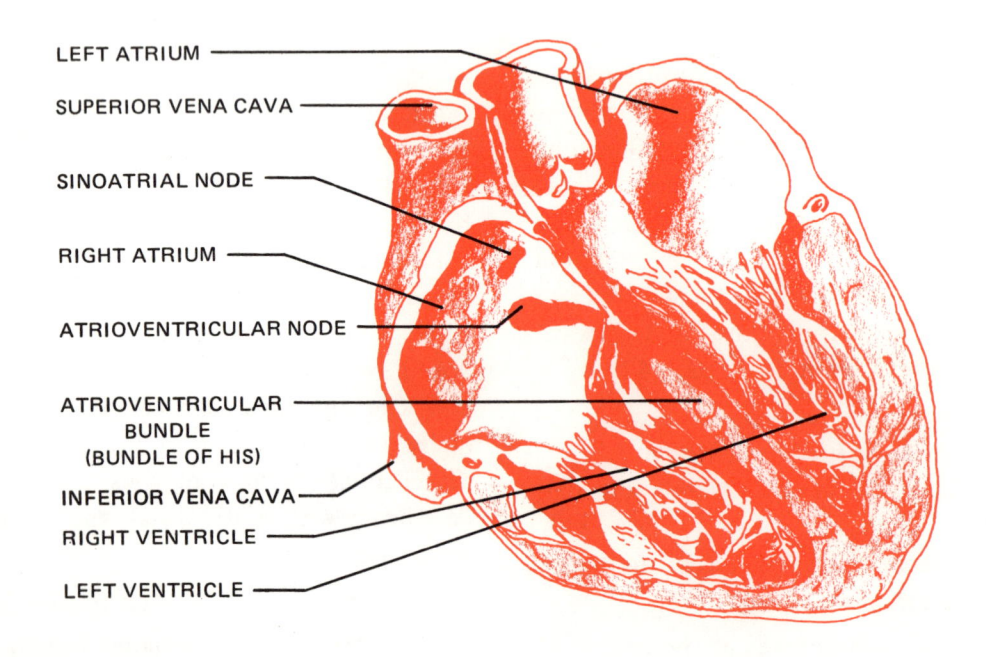

Fig. 14-2 Note location of the S-A and A-V nodes

THE HEART

open tricuspid and mitral valves. At the same time, the ventricles are relaxed, allowing them to fill with the blood from the atriums. At this point since the semilunar valves are closed, the blood cannot enter the pulmonary artery or the aorta.
2. The A-V node stimulates the contraction of both ventricles so that the blood in the ventricles is pumped into the pulmonary artery and the aorta through the semilunar valves which are now open. At this point the atriums are relaxed and the tricuspid and mitral valves are closed.
3. The ventricles relax, the semilunar valves are closed to prevent the blood flowing back into the ventricles, and the cycle begins again with the signal from the S-A node.

SUGGESTED ACTIVITIES

- If laboratory facilities are available, obtain a calf or sheep heart to examine. Notice the pericardium which is the tissue-like covering around the heart. Locate the atriums, ventricles, and identify the four valves.

- Locate and label the parts of the heart indicated on the diagram. Include valves, vessels, and nodes.

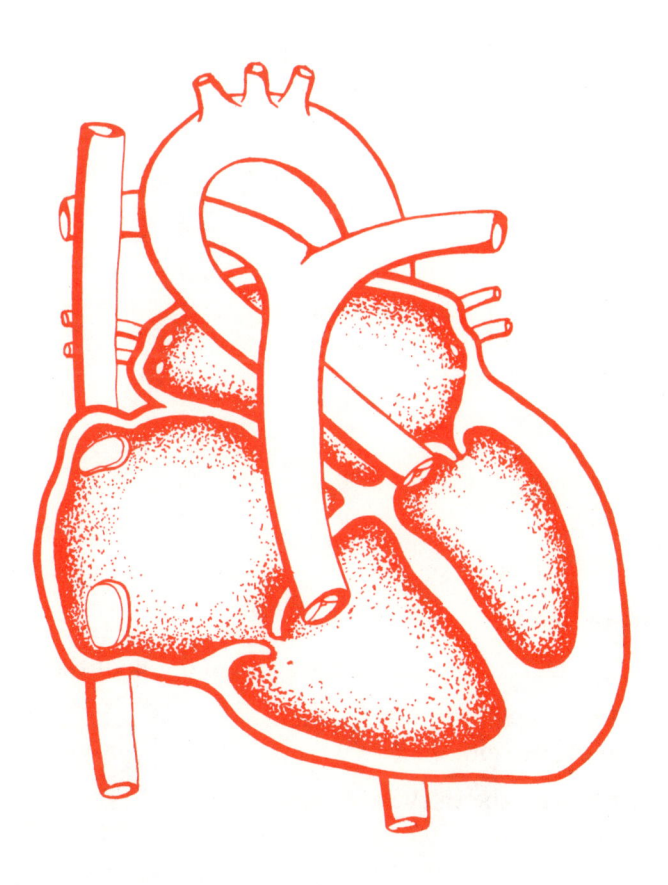

TRANSPORT OF FOOD AND OXYGEN

REVIEW

A. Briefly answer the following questions.

1. What is the pericardium? Describe its function.

2. Describe the functions of the endocardium.

3. Name the four chambers of the heart.

4. Name each valve and then locate its position by identifying the structure or area on both sides of the valve.

5. What function does a valve perform?

B. Complete the following statements.

1. The heart or cardiac muscle is called the _____.
2. The _____ is the smooth lining of the heart.
3. The structure known as the _____ stimulates the contraction of the ventricles.
4. The ventricle with the thickest muscular structure is on the _____ side of the heart.
5. The _____ is also called the pacemaker of the heart.

unit 15 function and path of general circulation

OBJECTIVES

Studying this unit should help the student:

- Trace the path of the general circulation.
- Describe the four main functions of the general circulation.
- Name some specialized circulatory systems.

The function of the general (systemic) circulation is fourfold: it circulates nutrients, oxygen, water, and secretions to the tissues and back to the heart; it carries products such as carbon dioxide and other dissolved wastes away from the tissues; it helps equalize body temperature; it aids in protecting the body from harmful bacteria.

The blood of the general circulation leaves the heart from the thick-walled left ventricle, by way of the aorta, the largest artery in the body. From the aorta, branches

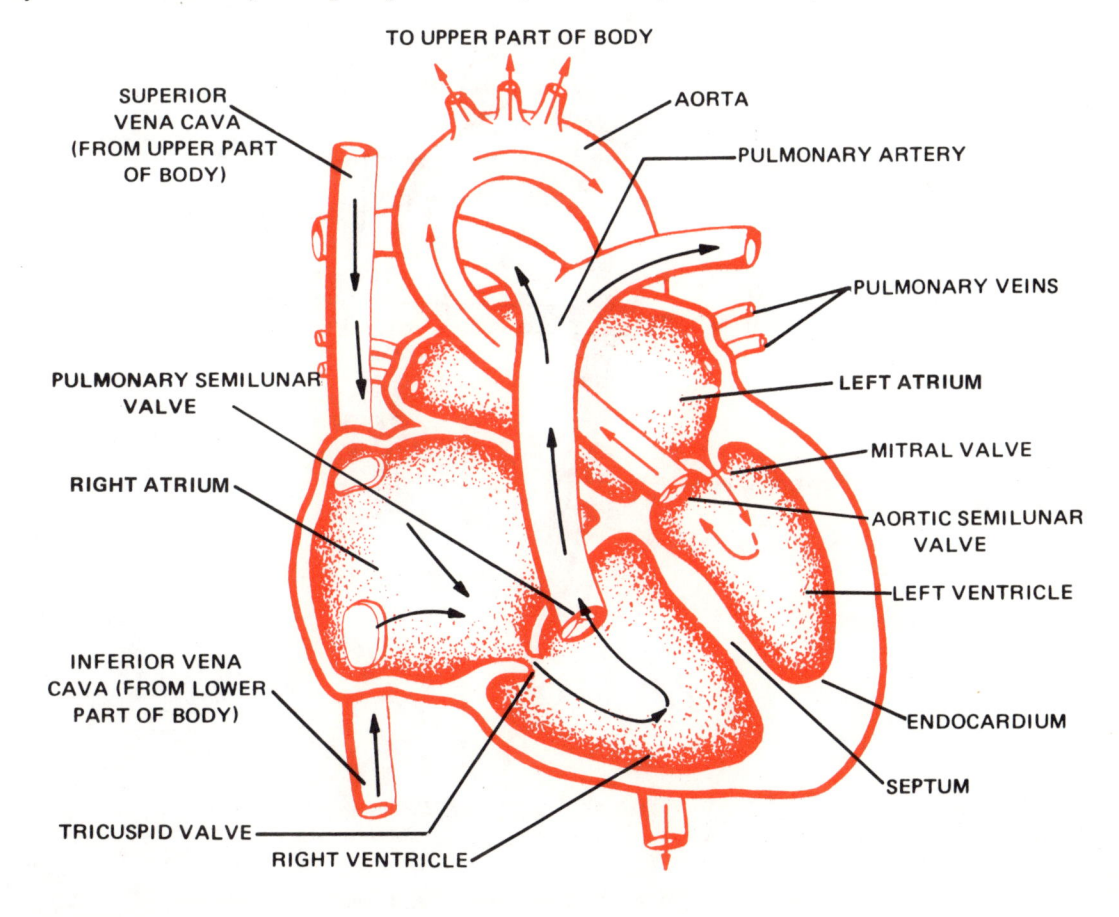

45

TRANSPORT OF FOOD AND OXYGEN

lead to the head, arms, trunk, and legs. When the circuit is completed, this blood returns to the opposite side of the heart from which it began. It enters the right side of the heart by way of the superior and inferior vena cava, figure 15-1, page 45.

Some circulatory routes within the general circulation are called by special names. The *coronary circulation,* which serves the heart, is a part of the general circulation. It has two branches, left and right. These branches come from the aorta just above the heart. The branches encircle the heart muscle, with many tiny branches going to all parts of the heart muscle. A special group of veins returns the blood to the right atrium.

The *renal circulation* is that part of the general circulation which carries blood from the aorta through the renal artery to the kidneys and then through the renal vein back to the heart by way of the inferior vena cava. This renal vein carries limited amounts of cell waste since many of them have been removed from the blood in the kidneys.

The *portal circulation* is a branch of the general circulation which includes veins from the spleen, pancreas, stomach, small intestine, and colon. These veins carry digested food plus water to the liver. Thus the *hepatic vein* which goes from the liver to the inferior vena cava is heavily supplied with food on its way to the body tissues to nourish them.

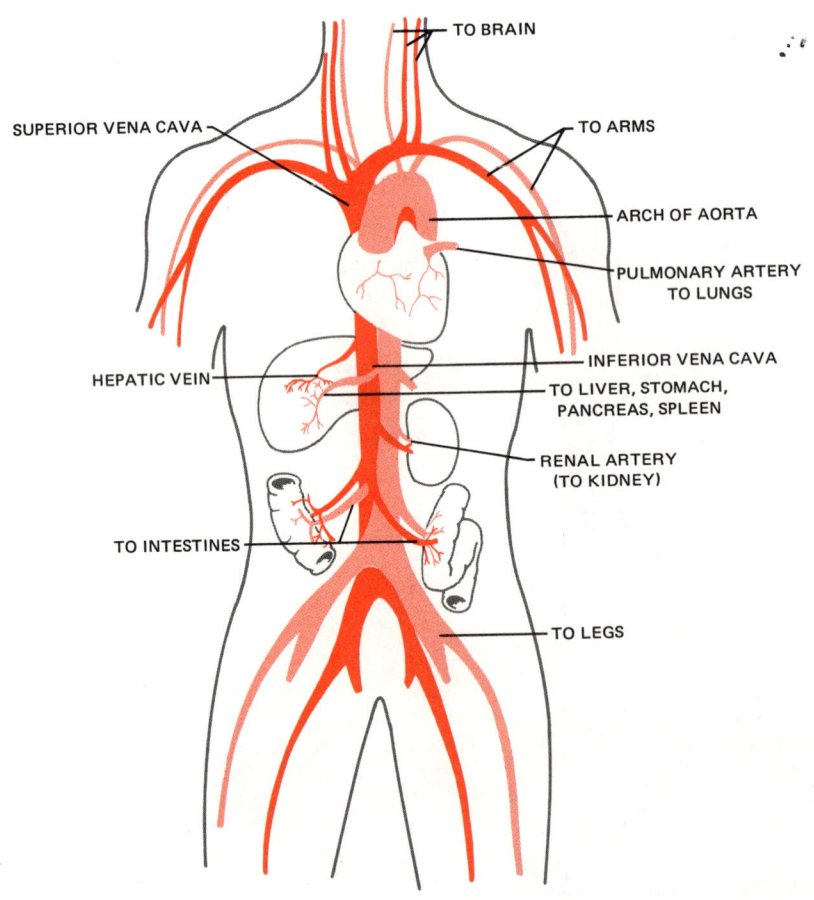

Fig. 15-2 General circulation showing arterial and venous distribution (arteries — light, veins — dark)

FUNCTION AND PATH OF GENERAL CIRCULATION

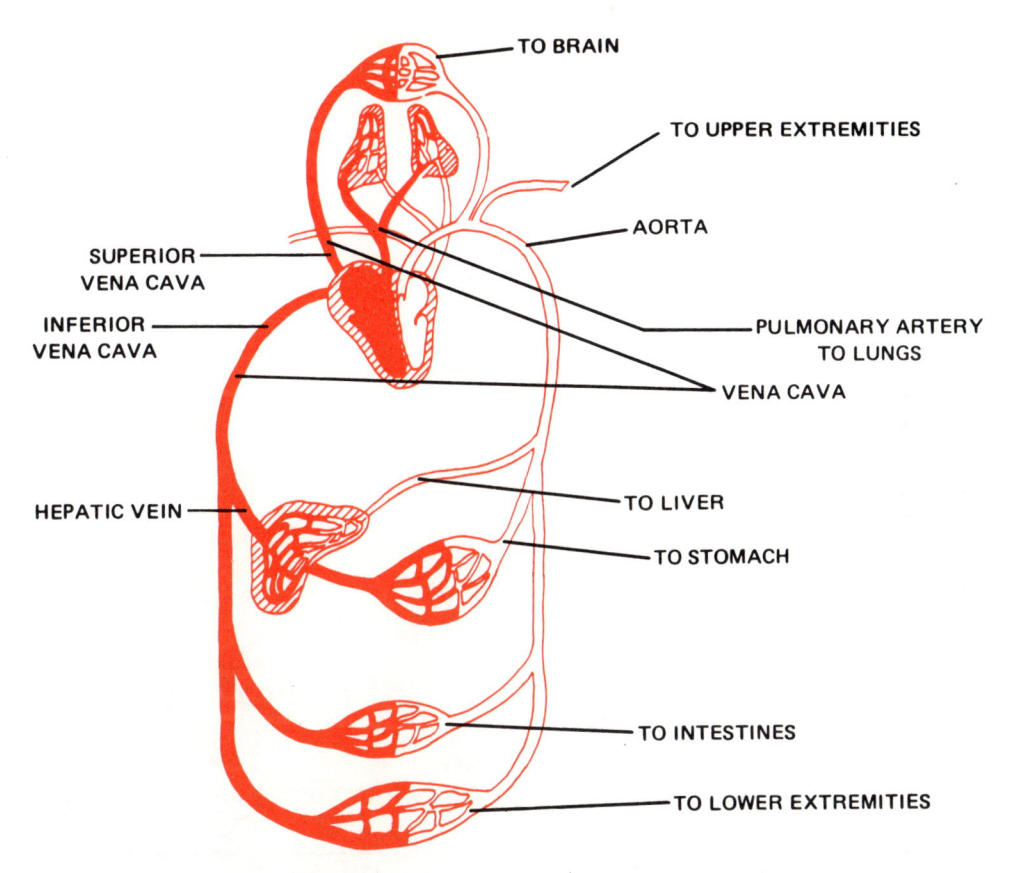

Fig. 15-3 Schematic diagram of the general circulation

SUGGESTED ACTIVITIES

- Using figure 15-3, trace the blood from the aorta to the vena cava.

REVIEW

A. Briefly answer the following questions.
1. Describe the chief functions of the general circulatory system.

TRANSPORT OF FOOD AND OXYGEN

2. Name the special circulatory system which is responsible for each of the following functions.

   a. Carries water to the kidneys

   b. Nourishes the heart

   c. Serves the digestive organs

B. Complete the following statements:
1. The blood leaves the heart from the _____ ventricle through the _____ , the largest artery in the body.
2. The deoxygenated blood re-enters the heart through the _____ _____ and into the _____ atrium.

C. Match each of the terms in column I with its correct statement in column II.

Column I	Column II
_____ 1. coronary circulation _____ 2. renal vein _____ 3. portal circulation _____ 4. renal circulation _____ 5. hepatic vein	a. a part of the general circulation which is made up of veins and carries digested food and water to the liver b. that part of the general circulation which carries blood from aorta to the kidneys and back to the heart c. the part of the general circulation which nourishes the heart d. the great vein of the general circulatory system e. carries limited amounts of cellular waste back to the heart by way of the inferior vena cava f. the vein which goes from the liver to the inferior vena cava heavily supplied with nutrients

unit 16 pulmonary circulation

OBJECTIVES

Studying this unit should help the student:

- Trace the route of the pulmonary circulation.
- Describe the function of the pulmonary circulation.
- Contrast the general and the pulmonary circulatory systems.

The pulmonary circulation carries blood from the heart to the lungs and back to the heart. The blood carried by the pulmonary artery is deoxygenated blood which is a darker red than when it leaves the lungs. There are also waste products in this blood. One of the waste products carried by the blood from the cells is carbon dioxide. It is exchanged for a new supply of oxygen in the lungs. The now oxygenated blood is bright red.

The pulmonary circulation starts its circuit by leaving the right ventricle of the heart through the pulmonary artery, figure 16-1. This artery carries the blood

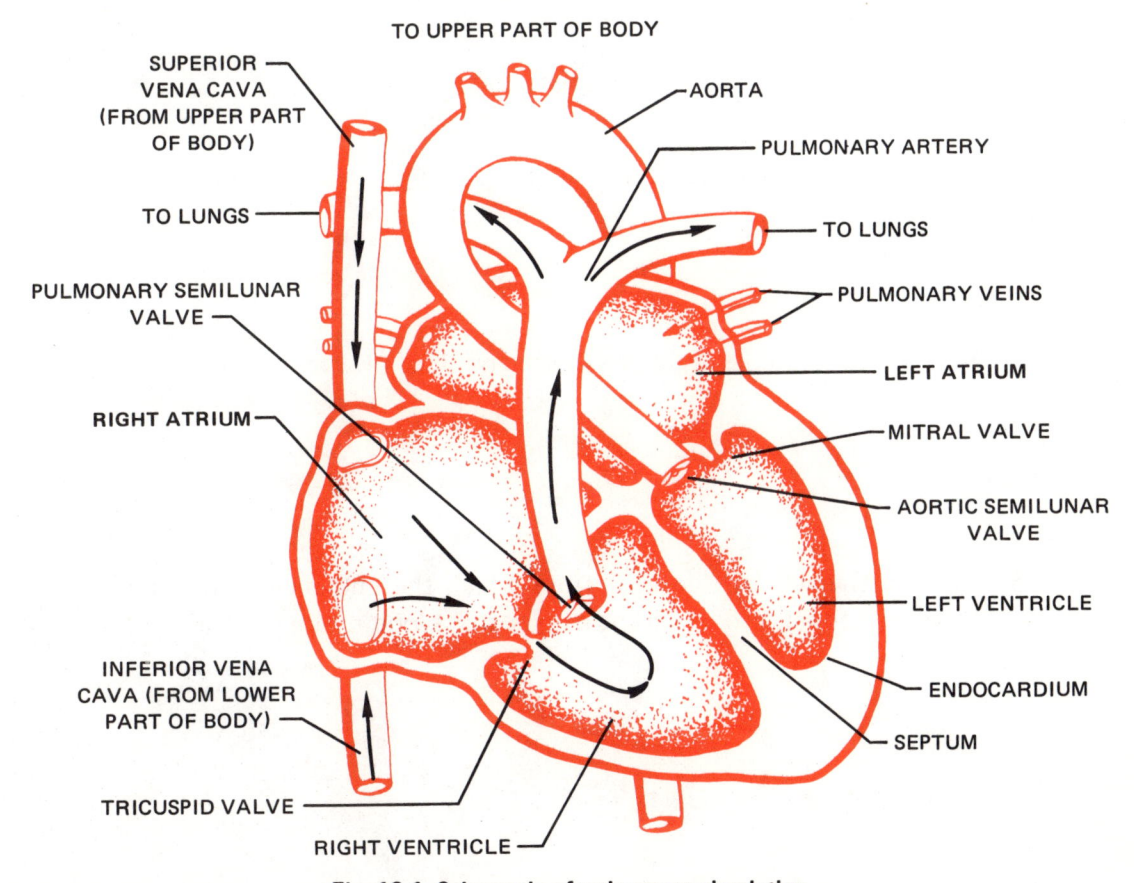

Fig. 16-1 Schematic of pulmonary circulation

TRANSPORT OF FOOD AND OXYGEN

to the capillaries of the lungs. Here the exchange of carbon dioxide for oxygen takes place. The blood, freshly supplied with oxygen, returns to the heart by way of the pulmonary veins. It enters the left atrium of the heart, the opposite side from which it left. It is now ready to make its circuit throughout the body by way of the general circulation to distribute a fresh supply of oxygen.

SUGGESTED ACTIVITIES

- Research and discuss what is meant by a "blue" baby. Why does the baby have a bluish appearance? What can be done for this baby to help it survive?

REVIEW

Briefly answer the following questions.

1. Which two main organs are involved in the pulmonary circulation?

2. Which pulmonary vessel carries deoxygenated blood?

3. Which major waste material is present in deoxygenated blood?

4. What function do the capillaries in the alveoli of the lungs perform?

5. Identify the side of the heart from which the pulmonary circulation leaves and the side to which the oxygenated blood returns?

unit 17 blood vessels

OBJECTIVES

Studying this unit should help the student:

- List the five types of blood vessels.
- Describe the particular functions of each type of blood vessel.
- Identify the principal blood vessels of the body.

The heart pumps the blood to all parts of the body through a remarkable system of three types of blood vessels: arteries, capillaries, and veins, figure 17-1.

Arteries carry blood away from the heart. They have elastic walls which expand when the surge of blood enters the artery after the contraction (pumping beat) of the heart. They then relax to normal size when the rush of blood pauses and the ventricles fill for the next contraction. The aorta, the artery leading away from the heart, is the largest blood vessel of the body, figure 17-2, page 52. The aorta leads to smaller arteries, then to tiny arteries called *arterioles*. The blood from the left ventricle follows this route and finally reaches the capillaries.

The *capillaries* are so small they must be seen through a microscope. They are barely large enough to allow the passage of the red blood cells. They connect the arterioles with tiny veins, called *venules*. The venules connect with larger veins. The capillary walls are extremely thin so that nutrients can pass out through them to the surrounding tissues and waste products from the tissues can pass back into the bloodstream in the same way. Tiny openings in the walls allow white blood cells to leave the bloodstream and enter the tissue spaces. In the capillaries, too, some of the plasma diffuses out of the blood and becomes tissue fluid. This is later returned to the blood through the lymphatic vessels.

The *veins* carry blood back to the heart. They are much less elastic than arteries. The pressure from the pumping action of the heart is much diminished by the time blood reaches the veins for its return journey. The veins have valves which allow the blood to flow only in the direction of the heart, thus preventing backflow, figure 17-3, page 52. The largest vein in the body is the vena cava; its

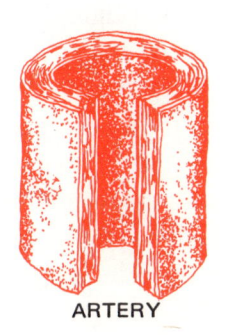

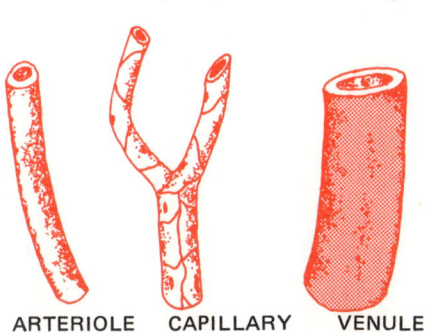

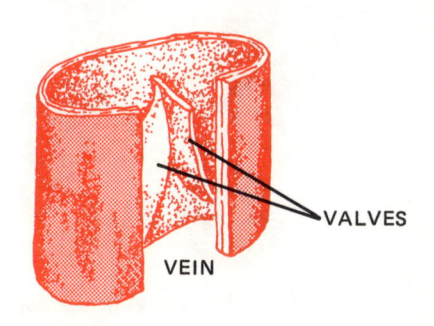

Fig. 17-1 Types of blood vessels

TRANSPORT OF FOOD AND OXYGEN

NAME	AREA SERVED
Common carotid Innominate Right and left coronary Lateral thoracic Pulmonary Aortic arch Right and left subclavian Thoracic aorta Right and left palmar digital Right and left ulnar Right and left radial Right and left brachial	Neck, head and chest
Abdominal aorta Celiac Superior mesenteric Renal	Abdominal region
Common iliac Right and left external iliac	Abdomen and legs
Right and left deep femoral Right and left femoral Right and left popliteal Right and left anterior tibial Right and left posterior tibial Right and left dorsal pedis	Legs

Fig. 17-2 Principal arteries

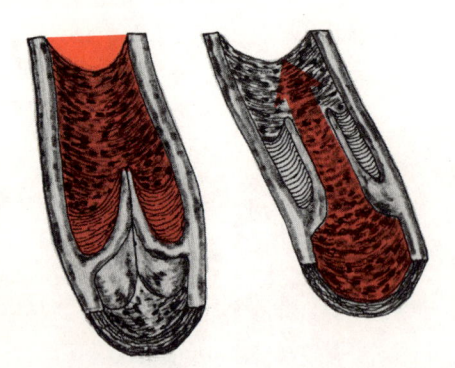

Fig. 17-3 Valves in the vein

superior and inferior branches enter the right atrium of the heart.

BLOOD PRESSURE

Initially, when the heart pumps blood into the arteries, the surge of blood filling the vessels creates pressure against their walls. The pressure at the moment of contraction is the *systolic blood pressure,* caused by the rush of blood which follows contraction of the ventricles. The lessened force of the blood (when the ventricles are relaxed) is called *diastolic pressure.* The pressure present in the arteries that are close to the initial surge of blood is greatest and gradually decreases as the blood travels further away from the pumping action.

BLOOD VESSELS

NAME	AREA SERVED
Internal jugular External jugular Right and left innominate Right and left subclavian Pulmonary Superior vena cava Inferior vena cava Right and left axillary	Head, Neck, Chest
Right and left cephalic Right and left basilic	Arms
Hepatic Portal Splenic Common iliac	Abdominal region
Right and left great saphenous Right and left femoral Right and left popliteal Right and left posterior tibial Right and left anterior tibial Right and left dorsal venous arch	Legs

Fig. 17-4 Principal veins

SUGGESTED ACTIVITIES

- Investigate and be prepared to discuss the following questions. What is blood pressure? Where is it usually taken? How is it measured? What instruments are used for taking blood pressure?
- Try the following experiment with a classmate and discuss the findings.
  Take the pulse for 30 seconds while the student is seated.
  Take the pulse for 30 seconds while the student is standing.
  Ask the student to jump up and down 25 times, being sure that the knees are well flexed.
  - a. Check the pulse immediately following exercise.
  - b. Check the pulse one minute after exercise.
  - c. Check the pulse two minutes after exercise.

REVIEW

Match each of the terms in column I with its correct statement in column II.

Column I	Column II
_____ 1. arteries _____ 2. capillaries _____ 3. valves _____ 4. veins _____ 5. arterioles	a. smaller arteries which lead to capillaries b. permit blood to flow in only one direction c. enters the right atrium of the heart d. blood vessels which carry blood back to the heart e. connects arterioles with venules f. large thick muscular walled blood vessels carrying blood away from the heart

unit 18 the blood

OBJECTIVES

Studying this unit should help the student:

- List the important parts of the blood.
- Describe the function of each part.
- Recognize the significance of the various blood types.

The average adult has four to five quarts of blood in his body. Loss of more than one quart at any one time leads to a serious condition.

The blood is the transporting fluid of the body. It carries nutrients from the digestive tract to the cells, oxygen from the lungs to the cells, waste products from the cells to the various organs of excretion, and hormones from secreting cells to other parts of the body. It also aids in the distribution of heat formed in the more active tissues (such as the skeletal muscles) to all parts of the body.

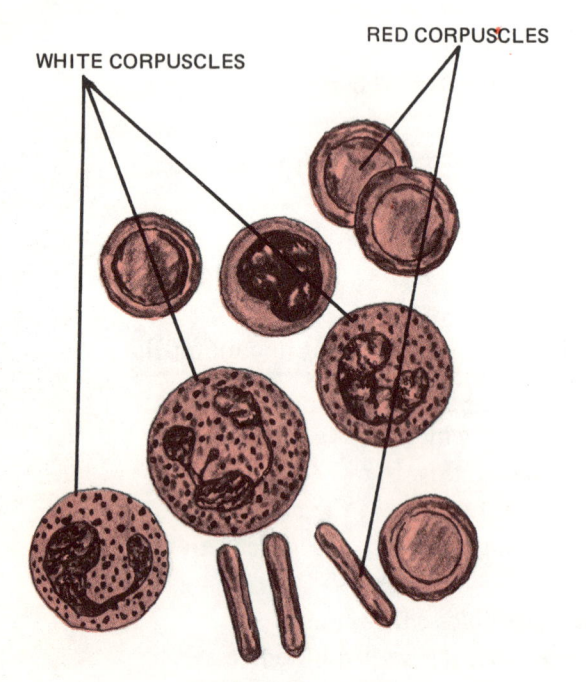

Fig. 18-1 Blood cells

BLOOD COMPOSITION

Blood is composed of a liquid portion called *plasma* and several different kinds of blood cells, figure 18-1. The plasma is a straw-colored liquid that comprises about 55 percent of the blood volume. It is composed primarily of water and a number of plasma proteins such as fibrinogen and prothrombin (active in blood clotting), albumins, enzymes, and globulins. The latter are the basis of blood types, and some of them act as defensive antibodies which destroy or render harmless various disease-causing organisms. Nutrients, mineral salts, and waste products are dissolved in the plasma, and the various kinds of blood cells float in it.

Blood cells may be grouped into three types: erythrocytes, leukocytes and thrombocytes. *Erythrocytes,* red blood cells, contain a red coloring substance called *hemoglobin.* The hemoglobin is the oxygen-carrying part of the blood. Erythrocytes also contain an antigen called the *Rh factor* which must be considered in prenatal care and in giving blood transfusions. The presence of the Rh factor is inherited. About 85 percent of Americans have it and are said to be Rh positive, while 15 percent do not have it and are said to be Rh negative.

Erythrocytes are made in the red bone marrow of essentially all bones up until adolescence. (In the fetus, red blood cells are also made in the spleen and liver.) As one grows older, the red marrow of the long bones is replaced by fat marrow; erythrocytes are then formed only in the short and flat bones. Erythrocytes have no nucleus and live only about 120 days. They are then broken down by the spleen and liver. The hemoglobin breaks down into *globin* and *heme*. The iron content of heme is used to make new red blood cells. The normal count ranges from 4,500,000 to 5,500,000 red blood cells per cubic millimeter of blood. The hemoglobin count is 14 to 16 grams per 100 cubic centimeters of blood.

Leukocytes are the white blood cells which help to protect the body against infections. One type of leukocyte, called a *phagocyte,* goes to the site of an infection to surround, engulf, and destroy germs. All leukocytes, unlike erythrocytes, can move through the intercellular spaces of the capillary wall. Most of them are made in the bone marrow, but some are made in lymphatic tissue. They vary in shape and size. Little is known about the life span or destruction of the white blood cells. A normal leukocyte count averages 5,000 to 9,000 cells per cubic millimeter of blood.

Thrombocytes, also called blood platelets, are plate-shaped cells which initiate the blood-clotting process. They are formed in the bone marrow and disintegrate in the bone marrow or lungs. The mode of disintegration is still in question. The normal blood platelet count is 250,000 to 450,000 for each cubic millimeter of blood.

BLOOD CLOTTING PROCESS

Blood clotting is a complicated and essential process which depends in large part on thrombocytes. When a cut or other injury ruptures a blood vessel, clotting must occur to stop the bleeding. On the other hand, unnecessary clotting can clog vessels, cutting off the vital supply of oxygen.

Although the exact details of this process are not clear, it is known that platelet factors induce the formation of thromboplastin. It is thought that the blood platelets rupture on contact with a rough surface and yield *thromboplastin*. The thromboplastin with some other blood factors forms *prothrombin*. Prothrombin, in the presence of calcium, becomes *thrombin*. Thrombin reacts with the soluble fibrinogen (of the plasma) to form insoluble *fibrin*. The fibrin is like many tangled threads which hold the blood cells and platelets, thus forming a clot.

Prothrombin supply, which is necessary to the blood-clotting process, is dependent on vitamin K. Vitamin K is a catalyst in the synthesis of prothrombin by liver cells. It is synthesized in the body by a type of bacteria found in the intestines. Some vitamin K may be found in the diet.

BLOOD TYPES

There are different kinds of blood groups called *types*. An individual inherits his blood type from his parents. Blood must be crossmatched as a safety measure for the patient before a transfusion is administered. The various blood types are shown in figure 18-2, page 56.

BLOOD NORMS

Tests have been devised to use physiological blood norms in diagnosing and following the course of certain diseases. Some of these norms are listed in figure 18-3, page 56.

As stated earlier, prothrombin is a factor found in blood plasma, which is needed for coagulation. The test to determine the prothrombin concentration in the blood plasma is made before and after the administration of vitamin K. If such concentration takes longer to appear than the time shown in figure 18-3, liver damage or failure to absorb vitamin K is suspected as a cause.

TRANSPORT OF FOOD AND OXYGEN

TYPE	PERCENT IN POPULATION	CROSSMATCH INFORMATION
O	44%	Universal donor can give blood to types A, AB, B, and O but can only receive type O.
AB	3%	Universal recipient can give only to type AB, but can receive blood from types A, B, AB, and O.
A	41%	Can give blood to types A and AB and can receive from types A and O.
B	12%	Can give to types B and AB and can receive from types B and O.

Fig. 18-2 Blood types

TEST	NORMAL RANGE
Bleeding time	1 to 3 minutes
Coagulation time	6 to 12 minutes
Hemoglobin count	14 to 16 gms per 100 cc.
Platelet count	250,000 to 450,000 per cubic millimeter
Prothrombin time (quick)	10 to 15 seconds
Sedimentation rate (Westergren) in first hour	Men — 0 to 12 millimeters Women — 0 to 20 millimeters

Fig. 18-3 Blood norm tests

Sedimentation rate is the time required for erythrocytes to settle to the bottom of an upright tube at room temperature. It indicates whether disease is present and is very valuable in observing the progression of inflammatory conditions. The Westergren method shows the normal rate for women slightly higher than for men.

SUGGESTED ACTIVITIES

- If laboratory facilities are available, practice determining hemoglobin count. Cleanse the tip of your finger with alcohol (70%). Prick it with a sterile needle. Place a large drop of blood on a piece of filter paper. As soon as the fluid is absorbed, compare it with the color scale which your instructor will have available. What is your hemoglobin? What is a normal count?

- Prepare a slide using a few drops of your own blood. Examine it under the microscope. Note the red and white blood cells.

- Before a transfusion may be given, blood must be typed and cross-matched. Discuss why this is necessary.

REVIEW

A. Briefly answer the following questions.

1. Name the three major types of blood cells.

2. What name is given to the straw-colored liquid portion of the blood?

3. What five proteins are contained in the blood?

4. Which part of the red blood cell is responsible for carrying oxygen?

5. Which area of the body is the primary site for red blood cell production?

6. Which type of blood cell protects the body against infection?

7. Which blood cell initiates the blood-clotting process?

8. Which substance in the clotting process finally acts to form the tangled threads of the clot?

9. Name six blood tests often relied upon to diagnose and follow the course of certain diseases.

10. Which two organs of the body break down red blood cells?

B. Select the letter which most correctly completes the statement.
1. The universal donor is
   a. type B
   b. type A
   c. type AB
   d. type O
2. The universal recipient is
   a. type B
   b. type A
   c. type AB
   d. type O
3. Negative Rh blood is found in
   a. 5% of the population
   b. 10% of the population
   c. 15% of the population
   d. 20% of the population
4. The type of blood found in the largest percent of the population is
   a. type O
   b. type A
   c. type AB
   d. type B
5. The prothrombin in the blood-clotting process is dependent upon
   a. vitamin A
   b. vitamin K
   c. vitamin P
   d. vitamin D

C. Select the most appropriate answer.
1. One of the following is not a blood cell.
   a. erythrocyte
   b. leukocyte
   c. neurocyte
   d. phagocyte
2. Erythrocytes contain all but one of the following elements.
   a. Rh factor
   b. leukocytes
   c. hemoglobin
   d. globin and heme
3. What characteristic is not true of normal thrombocytes?
   a. They average 4500 for each cubic millimeter of blood
   b. They are also called platelets
   c. They are plate-shaped cells
   d. They initiate the blood-clotting process
4. The normal leukocyte cell
   a. can only be produced in the lymphatic tissue
   b. goes to the site of infection where it engulfs and destroys germs
   c. is too large to move through the intracellular spaces of the capillary wall
   d. exist in numbers which amount to an average of 12,000 cells per cubic millimeter of blood
5. The blood-clotting process
   a. requires a normal platelet count which is 5,000 to 9,000 for each cubic millimeter of blood
   b. is delayed by the rupture of platelets which produces thromboplastin
   c. occurs in less time with persons having type O blood
   d. requires vitamin K for the synthesis of prothrombin

unit 19 lymphatic system

OBJECTIVES

Studying this unit should help the student:

- Describe the lymphatic system.
- Define the components of the lymphatic system.
- Outline the function of the lymph nodes.

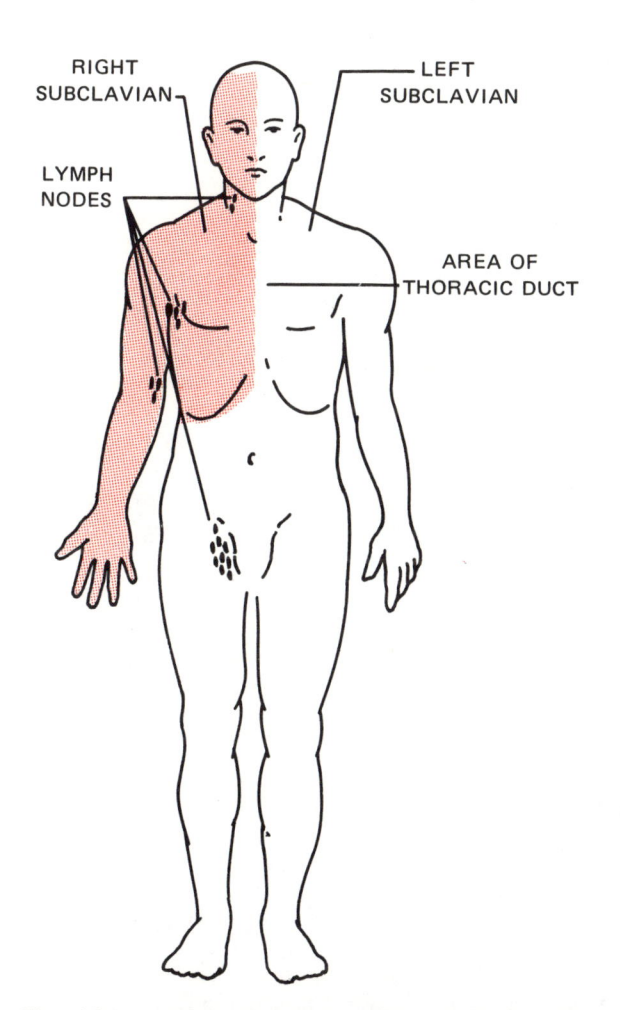

Fig. 19-1 Lymph drainage. Most of the lymph enters the circulation via the thoracic duct but the right lymphatic ducts drain lymph from the part of the body shown by the darkened area.

The *lymphatic system* consists of tiny lymph vessels, larger lymph vessels, and enlargements in the vessels called lymph nodes. It has no pump similar to the heart in the blood circulatory system. The lymph, the tissue fluid collected by the lymphatics, is kept moving by the contraction of any of the muscles in the surrounding tissues.

Lymph is a fluid formed from blood plasma. Some of the liquid diffuses through the walls of the capillaries into the tissue spaces. The lymph acts as a middleman between the blood and the tissues by carrying digested food and oxygen to the cells and picking up the waste products of cell metabolism; it is the fluid medium through which liquid and gaseous materials are exchanged between the body cells and the blood.

Most of the lymph re-enters the general circulation through the thoracic duct which empties into the left subclavian vein at the left shoulder, figure 19-1. The remainder of the lymph drains into the right subclavian vein. In this way, the waste products get back into the general circulation and are transported to the organs of excretion.

Lymph nodes vary in size and are located alone or in groups in various places along the lymph vessels throughout the

TRANSPORT OF FOOD AND OXYGEN

body. Their function is to make lymphocytes, a type of white cell, and to serve as a filter for screening out harmful substances such as bacteria or cancer cells from the lymph. If the harmful substances occur in such large quantities that they cannot be destroyed by the lymphocytes before the lymph node is injured, the node becomes inflamed. This causes a swelling in the lymph glands — a condition known as *adenitis*.

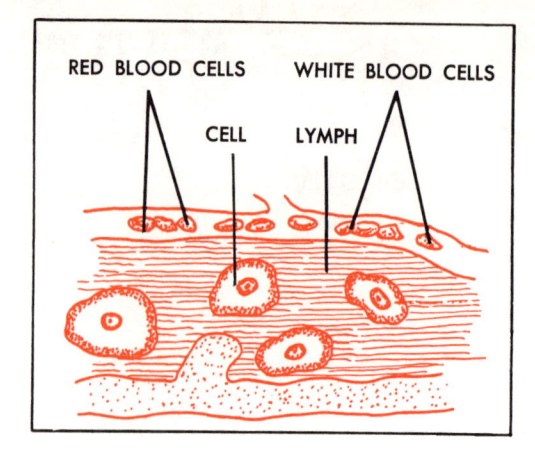

Fig. 19-2 Lymph circulation

SUGGESTED ACTIVITIES
- Discuss the diagram, figure 19-2. Explain how materials which the cell needs are transported.

REVIEW
Briefly answer the following questions.

1. From what substance is lymph formed?

2. What name is given to the enlarged portion of the lymph vessel?

3. The lymph allows exchange of digested food, oxygen and waste products between what two mediums of the body?

4. What body mechanism forces lymph to move through the lymphatic vessels?

5. Through which vessel does the lymph re-enter the general circulation?

6. Which two veins receive lymph drainage at the shoulder?

7. Name two functions of the lymph nodes.

8. Name the condition of inflammation which results in swelling of the lymph nodes.

9. Identify three body areas where lymph nodes are located.

10. Which of the three major types of blood cells may enter the lymph?

unit 20 disorders of the circulatory system

OBJECTIVES

Studying this unit should help the student:
- List disorders of the circulatory system.
- Describe some disorders common to the heart, blood vessels and the blood.
- Define new terms related to heart and blood disorders.

DISORDERS OF THE HEART

Acute rheumatic heart disease is an infection of the membrane lining of the heart, usually caused by a streptococcus organism.

Arrhythmia is any change or deviation from the normal orderly rhythm of the heart action.

Atrial fibrillation is a condition in which the atria are never completely emptied of blood. Their walls quiver instead of giving the usual contraction of a normal heartbeat.

Congenital heart disease is a condition in which the heart did not develop properly during fetal life.

Endocarditis is an inflammation of the membrane lining the heart.

Heart failure is the inability of the heart muscles to beat efficiently due to high blood pressure or other pathological conditions. This leads to lung congestion, dyspnea (difficult or rapid breathing) and frequent coughing.

Congestive heart failure is similar to heart failure but in addition there is edema (swelling) of the lower extremities. Blood backs up into the lung vessels and fluid extends into the air passages. The patient actually drowns in his own fluid.

Myocarditis is an inflammation of the heart muscle. *Myocardial infarction* is an area of the heart muscle which is damaged from lack of blood supply.

Pericarditis is an inflammation of the membrane covering the heart.

Angina pectoris is the severe chest pain which arises when the heart does not receive enough oxygen. It is not a disease in itself, but a symptom of an underlying problem with coronary circulation.

Murmurs may indicate some defect in the valves of the heart. They may take the form of gurgling and "hissing" sounds as the valves fail to close properly.

Coronary occlusion is a condition in which the heart does not receive enough blood because a coronary artery is blocked. This heart disease is also known as *coronary thrombosis* and may result from a *myocardial infarction*.

DISORDERS OF THE BLOOD VESSELS

Arteriosclerosis is a thickening of the walls of the arteries and loss of elasticity — "hardening of the arteries."

Gangrene is the death of body tissue due to an insufficient blood supply.

Phlebitis is an inflammation of the lining of a vein accompanied by clotting of blood in the vein.

Varicose veins are the swollen inelastic veins which result from the slowing up of blood flow back to the heart. Blood backs up in the veins if the muscles do not massage them. The weight of the stagnant blood distends the valves; the continued pooling of blood then causes distention and inelasticity of the vein walls.

Hemorrhoids are varicose veins in the walls of the rectum.

Cerebral hemorrhage is bleeding within the brain caused by arteriosclerosis or injury.

Aneurysm is a sac caused by enlargement of the blood vessel accompanied by thinning of the vessel wall.

DISORDERS OF THE BLOOD

Anemia is a deficiency of red blood cells or hemoglobin content. Primary anemia is caused by disease of the blood-forming organs.

Pernicious anemia is a condition in which there is an inadequate and abnormal formation of erythrocytes and the body is unable to absorb vitamin B_{12}.

Polycythemia is a condition in which too many red blood cells are formed.

Embolism is a clot which is carried by the bloodstream until it reaches an artery too small for passage.

Thrombosis is a blood clot which forms in a blood vessel and stays in the same place in which it was formed.

Hemophilia is a hereditary disease in which the blood clots slowly. This causes prolonged bleeding with even minor cuts or bumps.

Leukemia is a condition in which there is a great increase in the number of white blood cells. (Sometimes this is called cancer of the blood.)

Sickle cell anemia is a chronic blood disease which is inherited from both parents. The disease causes red blood cells to form in the abnormal crescent shape. These cells carry less oxygen and break easily causing anemia. The *sickling trait*, a less serious disease, occurs with inheritance from only one parent. Sickle cell anemia occurs almost exclusively in Negroes.

DISORDERS OF BLOOD PRESSURE

Hypertension is high blood pressure in which the systolic reading stays above 140 mm of mercury. The average adult reading depends on age, weight and build.

Hypotension is low blood pressure; usually the systolic reading is under 100 mm of mercury.

SUGGESTED ACTIVITIES

- Refer to reading references and outside sources to find out why phlebitis is dangerous.
- Discuss why patients with damaged hearts must restrict physical activity as recommended by their doctor.

REVIEW

Select the correct item which completes each statement.

1. Myocarditis is an inflammation of the
   a. lining of the heart
   b. covering of the heart
   c. arteries of the heart
   d. muscle of the heart
2. Leukemia is a condition in which there is a great increase in
   a. erythrocytes
   b. neurocytes
   c. fibrinogen
   d. leukocytes

TRANSPORT OF FOOD AND OXYGEN

3. Hypertension or high blood pressure is a condition in which the systolic reading stays above
    a. 190 mm of mercury
    b. 160 mm of mercury
    c. 140 mm of mercury
    d. 120 mm of mercury

4. Hemophilia is a condition in which the blood clots slowly and is acquired by
    a. environmental contacts and poor nutrition
    b. being inherited by the male but transmitted by the female
    c. close contact with persons who have the disorder
    d. contact with animals in the home and yard

5. A clot which is carried by the blood until it blocks a blood vessel too small for passage is called
    a. an embolism
    b. a thrombus
    c. an aneurysm
    d. a stenosis

section 4 transport of food and oxygen
SELF-EVALUATION

A. Match each of the terms in column I with its correct statement in column II.

Column I	Column II
____ 1. aorta	a. lower chambers of the heart
____ 2. atriums	b. white blood cells which absorb and destroy harmful bacteria
____ 3. cardiac	c. referring to the lungs
____ 4. coronary	d. largest artery in body
____ 5. endocardium	e. liquid part of blood
____ 6. hemoglobin	f. upper chambers of heart
____ 7. lymphatic	g. vessel transporting lymph
____ 8. phagocytes	h. circulation through kidneys
____ 9. pericardium	i. lining of heart
____ 10. portal circulation	j. arteries which nourish heart
____ 11. pulmonary	k. largest vein in body; returns to right atrium
____ 12. plasma	l. oxygen-carrying part of the blood
____ 13. renal	m. pertaining to the heart
____ 14. valves	n. goes to liver from small intestine
____ 15. vena cava (superior and inferior)	o. covering of heart
____ 16. ventricles	p. structures in heart and veins which permit the blood to flow in one direction only
	q. membrane lining the chest cavity

B. Encircle the letter before the correct answer.

1. An infection of the membrane lining the heart, which is usually caused by the streptococcus organism is
   a. acute pericarditis
   b. acute myocarditis
   c. acute rheumatic heart disease
   d. acute atrial fibrillation

2. Hypotension is a condition in which the systolic reading usually continues below
   a. 75 millimeters of mercury
   b. 100 millimeters of mercury
   c. 110 millimeters of mercury
   d. 120 millimeters of mercury

SELF-EVALUATION

3. A condition in which there are too many red blood cells is
   a. pernicious anemia
   b. leukemia
   c. polycythemia
   d. simple or primary anemia

4. A condition in which a blood clot may be carried by the bloodstream until it reaches a blood vessel too small for passage is a (an)
   a. embolism
   b. thrombus
   c. aneurysm
   d. stenosis

5. A condition in which there is an inability to absorb vitamin B_{12} is known as
   a. simple anemia
   b. pernicious anemia
   c. leukemia
   d. polycythemia

C. Match each of the terms in column I with its correct statement in column II.

Column I	Column II
____ 1. adenitis	a. varicose veins in walls of rectum
____ 2. anemia	b. blood clot in artery which nourishes the heart
____ 3. angina pectoris	
____ 4. arteriosclerosis	c. inflammation of a vein
____ 5. coronary thrombosis	d. moving blood clot in bloodstream
	e. inflammation of membrane which lines the heart
____ 6. embolism	
____ 7. endocarditis	f. hardening of the arteries
____ 8. hemorrhoids	g. severe chest pain
____ 9. myocarditis	h. deficiency of red blood cells
____ 10. phlebitis	i. inflammation of heart muscle
	j. inflammation of lymph glands
	k. inflammation of membrane that covers the heart
	l. crescent-shaped red blood cells

SECTION 5

BREATHING PROCESSES

unit 21 introduction to the respiratory system

OBJECTIVES

Studying this unit should help the student:

- Describe the function of the respiratory system.
- List the two stages of respiration.
- Describe the mechanics of respiration.

The respiratory system is composed of the organs having to do with bringing oxygen into the body and removing carbon dioxide from it. The term, *respiration*, has commonly been used to refer only to the inhalation of oxygen and the exhalation of carbon dioxide. This particular process, more properly called breathing, takes place in the lungs. True respiration, however, includes — in addition to the process of gas exchange — the use of the oxygen for energy release from food, and this takes place in the cells. This latter process is called *oxidation*, or burning. Just as wood when burned (oxidized) gives off energy in the form of heat and light, so does food give off energy when it is burned, or oxidized, in the cells. Much of this energy is released in the form of heat to maintain body temperature. Some of it, however, is used directly by the cells for such work as contraction of muscle cells, as well as to carry on other vital processes.

As wood burns, the carbon and hydrogen compounds that make up its substance are broken down and the carbon and hydrogen combine with oxygen to form carbon dioxide (CO_2) and water vapor (H_2O). Similarly, food, when oxidized, gives off waste products including carbon dioxide and water vapor. These are transported from the cells through the circulatory system to the lungs where they are exhaled. Respiration, therefore, occurs in two stages: external respiration and internal respiration.

External respiration is the exchange of oxygen and carbon dioxide between the body and the outside environment. It consists of inhalation and exhalation. As a person inhales, the air is warmed, moistened, and filtered. As he exhales, he gives off much of the carbon dioxide in the blood through the alveoli (air sacs) of the lungs. Some water vapor is also given off at this time.

Internal respiration is the exchange of carbon dioxide and oxygen between the cells and the lymph surrounding them, plus the oxidative process of energy in the cells. The differences in concentration of carbon dioxide and oxygen govern the exchange which occurs among the air in the alveoli, the blood and the tissue cells. The alveoli (after inhalation) are rich with oxygen and transfer the oxygen into the blood. The resulting greater concentration of oxygen in the blood, moves the oxygen into the tissue cells. Through respiration, the tissue cells use up the oxygen

BREATHING PROCESSES

and simultaneously build up a higher carbon dioxide concentration. This concentration increases to a point which exceeds the level in the blood. This causes the carbon dioxide to diffuse out of the cells and into the blood where it is then carried away to be eliminated.

SUGGESTED ACTIVITIES

- Breathe on a mirror. Note the moisture which appears on the mirror from the exhaled air. Discuss the fact that carbon dioxide, heat and water vapor are given off in exhalation.
- Jog in place. Note the effect of body activity on the rate of breathing. How does exercise change breathing? Why?
- Why does a young child breathe more rapidly than an aged person?

REVIEW

A. Briefly answer the following questions.

1. Name and describe the two stages of respiration.

2. Describe the oxidation process. What happens to the energy and waste products released?

B. Match each of the terms in column I with its correct statement in column II.

Column I	Column II
_____ 1. external respiration	a. exchange of gases between the cells and lymph surrounding them and energy release in cells
_____ 2. oxidation	b. carbon dioxide and water vapor
_____ 3. waste products of oxidation	c. air sacs in lungs
_____ 4. alveoli	d. filtering of air
_____ 5. internal respiration	e. combining of food and oxygen in tissues
	f. the inhalation of oxygen and exhalation of carbon dioxide
	g. water vapor only

unit 22 respiratory organs and structures

OBJECTIVES

Studying this unit should help the student:
- List the organs used in breathing.
- Define the structures within the lungs.
- Describe functions of parts of the respiratory system.

Figure 22-1 outlines the passageway through which air moves into the lungs. It includes the following structures:
- the nasal cavity (lined with cilia)
- the pharynx
- the larynx
- the trachea
- two bronchi
- bronchioles
- alveoli

The trachea divides to form the right and left bronchus. Each of these bronchi divide into

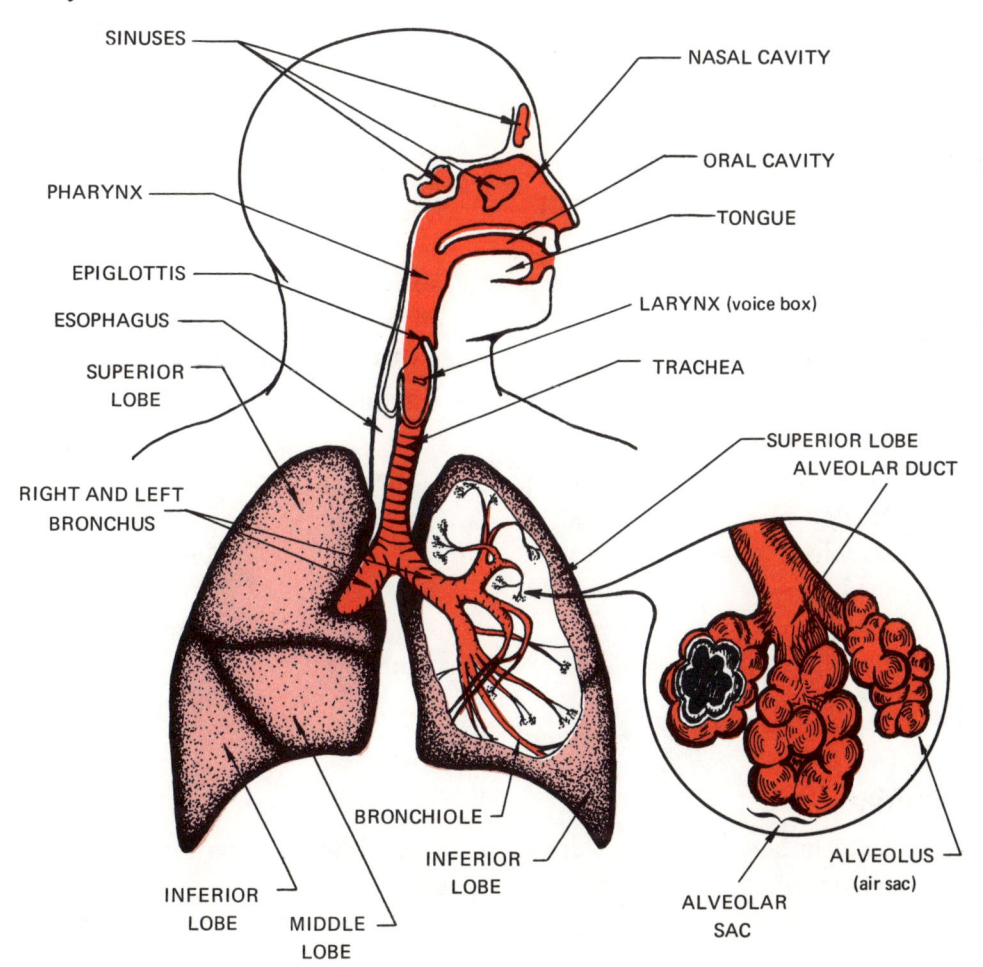

69

BREATHING PROCESSES

bronchioles which lead to the alveoli (air sacs). The diaphragm and the intercostal muscles support and aid the breathing process by bringing about the breathing movements.

The *alveolar sacs* consist of many alveoli and are made of a single layer of epithelial tissue. The alveoli are encased by a network of capillaries. The thin moist walls of both the alveoli and the capillaries permit rapid exchange of oxygen and carbon dioxide.

The *chest,* or *thoracic* cavity encloses the lungs. The chest has a double lining called the *pleura.* One layer of the pleura covers the lungs, and the other lines the thoracic cavity. Fluid between the double lining is present to prevent friction when breathing movements take place.

The upper respiratory tract is the upper terminal of the passageway to the lungs. In the nasal region a *nasal septum* divides the nasal chamber into left and right sides. This is made of cartilage but becomes bone as it nears the skull. Each nasal cavity is divided into three narrow passages by three scroll-like bony processes called the *turbinates.* All parts of the nasal cavities are covered with a moist mucous membrane. When air passes through the narrow passages it is warmed, moistened, and filtered. Nerve endings for the sense of smell are located in the mucous membrane in the upper part of the nasal cavity.

The *sinuses,* named frontal, maxillary, sphenoid, and ethmoid, are cavities of the skull in and around the nasal region. Short ducts connect the sinuses with the nasal cavity. Mucous membrane lines the sinuses and helps to warm and moisten air passing through. The sinuses also give resonance to the voice. The unpleasant voice sound of a nasal cold results from the blockage of sinuses.

SUGGESTED ACTIVITIES

- Using a plastic model or charts, trace the air passage from the nostrils to the air sacs of the lungs.
- If laboratory facilities are available, examine beef lungs and trace the trachea and bronchi into the tissues of the lung. Blow into the trachea with a glass tube and inflate the lungs. Observe the action.
- Discuss the effect lung congestion (as during a chest cold) has on breathing.
- Look up and learn the meaning of the following terms.

epiglottis	pleura	larynx
alveoli	respiratory center	bronchiole
medulla	mediastinum	mucosa
intercostals	nares	diaphragm
pharynx	ciliated mucous membrane	trachea
bronchi	apex	inspiration
base	expiration	

REVIEW

A. Explain how the air sacs are particularly adapted to permit a rapid exchange of oxygen and carbon dioxide.

B. Match each of the terms in column I with its correct statement in column II.

Column I	Column II
_____ 1. alveoli	a. help give resonance to voice
_____ 2. ciliated epithelium	b. air sacs
	c. lines nasal passage
_____ 3. intercostal muscles	d. voice box
	e. divisions of lungs
_____ 4. larynx	f. double lining of the thoracic cavity
_____ 5. lobes	
_____ 6. mucous membrane	g. chest
	h. collects dust particles
_____ 7. nasal septum	i. made of cartilage
_____ 8. pleura	j. support and aid the breathing process
_____ 9. sinuses	k. respiratory center
_____ 10. thoracic cavity	

C. Briefly answer the following questions.
1. What tissue makes up the alveolar sac?

2. What structure forms a tight network around the alveolar sac?

3. What structure in both the nasal passage and sinuses warms, moistens and filters air passing through?

4. Name the four sinuses in the nasal area.

5. Name the two muscles which are primarily responsible for the breathing movements.

unit 23 mechanics of breathing

OBJECTIVES

Studying this unit should help the student:

- Explain how breathing movements are controlled.
- Describe the action of the diaphragm and the ribs in breathing.
- Define the various lung capacities.

Ventilation of the lungs is due to changes in pressure which occur within the chest cavity. This variation in pressure is brought about by cellular respiration and mechanical breathing movements.

In the process of inhalation, the ribs are raised by contraction of the intercostal muscles. This increases the size of the chest cavity. At the same time, the diaphragm contracts and becomes flattened, moving downward. This serves to increase the space in the chest cavity in a vertical direction causing a decrease in pressure. Since atmospheric pressure is now greater, air rushes in, resulting in inhalation.

In exhalation, just the opposite takes place. The intercostal muscles and diaphragm relax. The ribs move down and the diaphragm moves upwards. These two movements decrease the space within the chest cavity, thus increasing the pressure. This increased pressure forces the air out, resulting in exhalation.

Breathing movements are controlled by the respiratory center in the brain; this is known as the *medulla*. This center sends rhythmic impulses to the diaphragm and rib muscles, causing them to contract or relax. Inspiration and expiration normally occur from sixteen to twenty-four times a minute depending on the activity, position, and age of the person. The respiratory center is affected by changes in the chemistry of the blood, particularly an increase in carbon dioxide content. This increase causes a corresponding increase in respiration rate while a decreased amount of carbon dioxide causes a decreased respiration rate.

Tidal volume (TV) is the amount of air that is inhaled and exhaled during rest. Normal tidal volume is about 500 milliliters.

Inspiratory reserve volume (IRV) is the extra volume of air that can be inhaled over and beyond the tidal volume. It is usually about 3000 ml.

The *expiratory reserve volume* (ERV) is the amount of air that can be forced by expiration after the end of a normal exhalation. This is about 1100 ml.

The *residual volume* (RV) is what remains in the lungs even after a forced expiration; this is about 1200 ml.

Vital capacity refers to the amount of air one can forcibly expire after a maximum inhalation; in other words, it is a measure of the ability to inspire and expire air. Disease processes that weaken the respiratory muscles (such as polio) or decrease the ability of the lungs to expand (emphysema) can decrease the vital capacity. The vital capacity of an average person is about 4500 ml.

SUGGESTED ACTIVITIES

- Demonstrate breathing movements by placing the hands near the base of the ribs on both sides of the chest. Note how the ribs rise and fall during breathing. Then place the hands horizontally across the stomach (at the base of the diaphragm) as inhalation and exhalation take place.

- Discuss the effects of the following on healthful breathing: posture, exercise, deep breathing, mouth breathing, enlarged tonsils and adenoids, tight clothing, smoking.

- Using library sources, look up the "iron lung." How does its action compare with natural breathing movements?

REVIEW

A. Briefly answer the following question.

1. Explain the function of

   (a) the diaphragm

   (b) the intercostal muscles

B. Match each of the terms in column I with its correct statement in column II.

Column I	Column II
_____ 1. respiratory control center	a. opposite of inhalation
_____ 2. inspiration and expiration	b. measure of the ability to inspire and expire air
_____ 3. vital capacity	c. complemental air
_____ 4. exhalation	d. located in the medulla
_____ 5. increases respiratory rate	e. occur from 16 to 24 times a minute
_____ 6. diaphragm	f. result of increase in carbon dioxide content of the blood
_____ 7. intercostal muscles	g. becomes flattened and moves downward during inhalation
_____ 8. tidal air	h. air which cannot be forcibly expelled from the lungs
_____ 9. residual air	i. corrresponds with atmospheric pressure
_____ 10. chest cavity space	j. air inhaled and exhaled during rest
	k. muscles in between the ribs which contract during inhalation
	l. moves upward during inhalation

unit 24 respiratory disorders

OBJECTIVES

Studying this unit should help the student:

- Describe some common respiratory diseases caused by a virus or bacteria.
- Describe some respiratory disorders unrelated to infectious causes.
- Suggest nursing care for respiratory ailments.

The greatest loss in manhours each year is caused by the common cold. This respiratory infection spreads quickly through the classroom, factory, or business office. It is also very often the basis for more serious respiratory disease. It lowers body resistance, making it subject to infection. The direct cause of a cold is usually a virus. The indirect or helping causes include chilling, fatigue, lack of proper food, and not enough sleep. A person who has a cold should stay in bed, drink warm liquids and fruit juice, and eat wholesome, nourishing foods.

The respiratory system is subject to various infections and inflammations caused by bacteria, viruses, and sometimes irritants. Many of these infections and inflammations follow the common cold.

Infectious Causes

Respiratory ailments due to infectious causes may affect the respiratory system:

Pharyngitis, is a red, inflamed throat which may be caused by one or several bacteria or viruses. It also occurs as a result of irritants such as too much smoking or speaking.

Laryngitis is an inflammation of the larynx, or voice box, and is often secondary to other upper respiratory infection. It can be recognized by the incidence of hoarseness or loss of voice.

Tonsillitis is an infection of the tonsils caused by one of several bacteria. Frequent occurrence of this infection, which is accompanied by severe sore throat, difficulty in swallowing, elevation of temperature, chills, and aching muscles, may result in surgical removal of the tonsils.

Sinusitis is an infection of the mucous membrane which lines the sinus cavities. One or several of the cavities may be infected. Pain and nasal discharge are symptoms of this infection, which, if severe, may lead to more serious complications.

Pleurisy is the result of inflammation of the pleura; it is often a complication of a more severe illness such as pneumonia or tuberculosis.

Bronchitis affects the mucous membrane of the trachea and the bronchial tubes. It may be acute or chronic and often follows infections of the upper respiratory tract.

Influenza is an infectious disease of the respiratory system which is caused by an identifiable virus. The symptoms cause pain and discomfort to the patient but are not usually serious. However, bronchopneumonia often follows the "flu."

RESPIRATORY DISORDERS

Pneumonia is an infection of the lung. The alveoli fill up with an inflammatory exudate. Pneumonia is usually caused by bacteria although there may be other causes. Onset is often sudden and is marked by chills and chest pain. When caused by a virus, it is called *viral* or *atypical* pneumonia.

Tuberculosis is an infectious disease of the lungs caused by the tubercle bacillus. Since there are no obvious early signs of this infection, yearly checkups with a chest X ray are important to detect its presence.

Diphtheria is a very infectious disease caused by an identifiable bacterium. The disease affects the upper respiratory tract and is recognized by the grayish-white or yellow membrane which forms on the tonsils and pharynx.

Noninfectious Causes

Respiratory ailments which are unrelated to infectious causes sometimes develop in the respiratory system:

Rhinitis is the inflammation of the nasal mucous membrane causing swelling and increased secretions.

Asthma is a respiratory disorder with difficult breathing, wheezing, coughing, presence of a mucoid sputum and a feeling of tightness in the chest. Serious episodes of wheezing may occur from emotional stress or from breathing irritants. Emergency care may be needed.

Atelectasis is a condition in which the lungs fail to expand normally.

Bronchiectasis is the dilatation of a bronchus along with heavy pus secretion.

Silicosis or miner's disease is caused by breathing silicone dioxide over a long period of time. The lungs become fibrosed which results in a reduced capacity for expansion.

Emphysema is a noninfectious condition in which the alveoli become overextended, with a resulting overinflation of the lungs. Respiration becomes increasingly difficult. Treatment is aimed at relieving the discomfort of the symptoms; at present, there is no cure.

Cancer of the lung is a malignant tumor which often forms in the bronchial epithelium. The incidence of this condition is high in middleaged men, especially those who are regular smokers. Early diagnosis is difficult, depending on examinations rather than apparent symptoms. Surgery is often indicated and if done soon enough is often successful.

Cancer of the larynx is curable if early detection is possible. It is found most frequently in men over fifty.

Nursing care in many of these respiratory ailments is directed toward making the patient as comfortable as possible and seeing that he gets sufficient rest and proper nourishment.

SUGGESTED ACTIVITIES

- Explain why sinusitis is more common among people who live at sea level. Explain why the climate in Arizona is usually beneficial to people who have sinusitis.

- Discuss how a minor infection can lower the body's resistance to other infections.

- Discuss how public health measures could help reduce the incidence of emphysema and lung cancer.

REVIEW

1. Describe what should be done for a person who is developing the symptoms of a cold.

2. Name five common respiratory disorders caused by a virus or bacterium.

3. Name three respiratory disorders of a noninfectious nature.

section 5 breathing processes
SELF-EVALUATION

A. Describe the process of external and internal respiration.

   (1) External respiration

   (2) Internal respiration

B. Explain the function of the hemoglobin in internal respiration.

C. Label the following drawing.

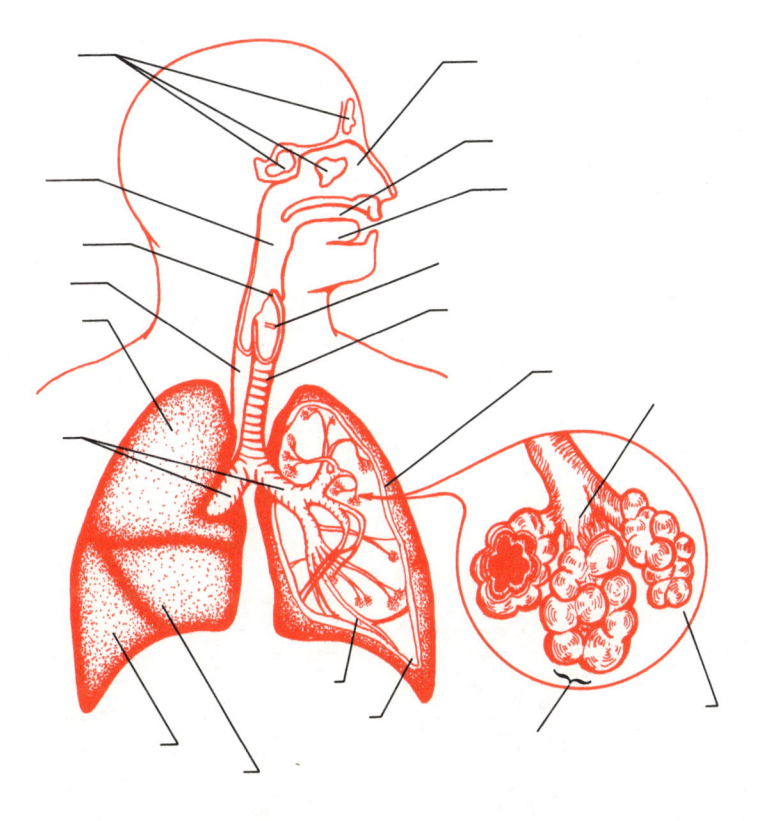

SELF-EVALUATION

D. Complete the following statements.

1. During inhalation, the ribs are _____ by contraction of the rib muscles.
2. The microscopic air sacs in the lungs are called _____ .
3. Laryngitis is an inflammation of the _____ .
4. When the diaphragm moves _____ , inhalation takes place.
5. Energy and waste products are given off when food unites with _____ .
6. The principal waste product of exhalation is _____ .
7. The respiratory system is dependent upon the _____ system for transporting oxygen to the body cells and carbon dioxide away from the body cells.
8. The lining of the thoracic cavity is called _____ .
9. The most common respiratory ailment is _____ .
10. The sinuses are lined with _____ .

E. Explain how the action of the respiratory center affects the breathing rate.

SECTION 6

DIGESTION OF FOOD

unit 25 introduction to the digestive system

OBJECTIVES

Studying this unit should help the student:

- Describe the general function of the digestive system.
- List the structures of the digestive system.
- Relate the function of the mouth and teeth to digestion.

All food which is eaten must be changed into a soluble, absorbable form within the body before it can be used by the cells. This means that certain physical and chemical changes must take place to change the complex food molecules into simpler soluble ones which can be transported by the blood to the cells and be absorbed through the cell membranes. The process of changing complex solid foods into simpler soluble forms which can be absorbed by the body cells is called *digestion*. It is accomplished by various digestive juices containing enzymes. *Enzymes* are substances that promote chemical reactions in living things.

The structures concerned with digestion make up the digestive system and compose the alimentary canal and accessory organs, figure 25-1. The *alimentary canal* includes the mouth, pharynx, esophagus, stomach, small intestine, and large intestine. It is a thirty- to forty-foot canal through which food passes during digestion. The *accessory* organs are the teeth, tongue, salivary glands, pancreas, liver and gallbladder.

THE MOUTH AND DIGESTION

The role of the mouth in the digestive process is to break up food into smaller particles by chewing. The *salivary glands*, which open into the mouth, manufacture an enzyme, *ptyalin*, which helps to change starch to sugar as a preliminary step in digestion. Saliva also moistens the food and thus aids swallowing. The study of digestion can begin with an examination of the structures of the mouth, figure 25-2.

The inside of the mouth is covered with mucous membrane. The tongue, lying on the floor of the mouth, assists in chewing and swallowing food. On the surface of the tongue are projections called *papillae,* some of which contain taste buds which respond to salt, sweet, bitter, and sour tastes.

Some of the salivary glands are located beneath the tongue on the floor of the mouth and beneath the mucous membrane lining the inside of the cheeks. A third location is the parotid glands, found on both sides of the face below the ear. Secretion by the salivary glands is stimulated by the sight of food and by its presence in the mouth, as well as by its taste and smell. Secretion of saliva, necessary to digestion, increases during the eating period.

The *hard palate* is a bony structure which forms the roof of the mouth. The

DIGESTION OF FOOD

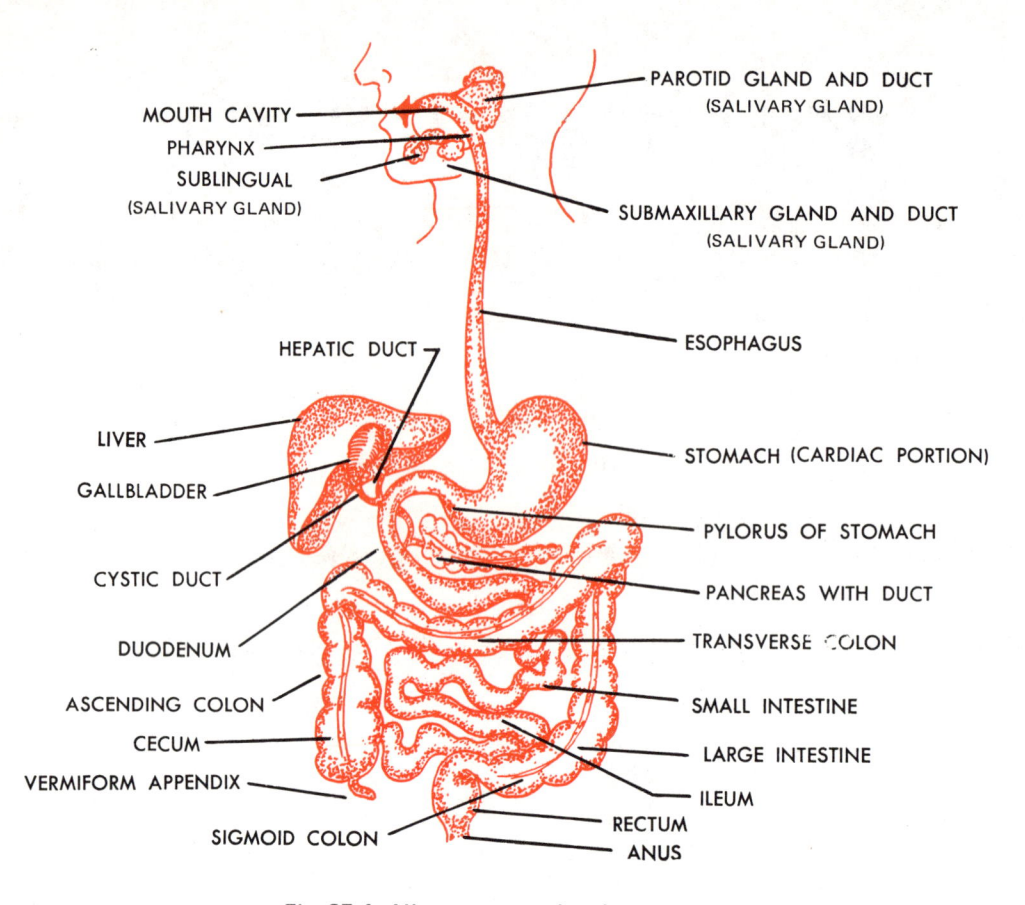

Fig. 25-1 Alimentary canal and accessory organs

soft palate is a curtain-like muscular fold of membrane adjoining the hard palate in the back of the mouth. A small, soft structure called the *uvula* is suspended from the center edge of the soft palate. Its function is not known.

The soft palate separates the mouth cavity from the pharynx. The *tonsils* are two prominent masses of lymphatic tissue, located on either side of the mouth cavity.

The *gingivae,* or gums, support and protect the teeth. They are made up of fleshy tissue covered with mucous membrane; this membrane surrounds the necks of the teeth and covers the upper and lower jaws.

The teeth are the hardest structure in the body since they are composed largely of mineral salts of calcium and phosphorus. The adult has thirty-two teeth which include

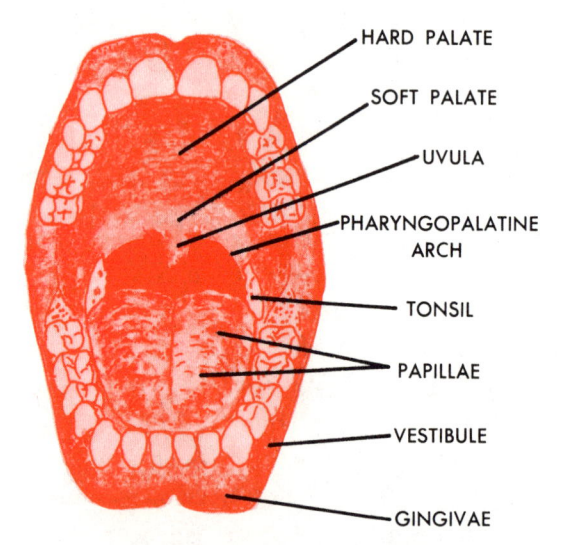

Fig. 25-2 Structure of the mouth

incisors for biting food, *canines* for tearing it, *bicuspids* and *molars* for crushing and chewing, figure 25-3, page 81. The shape

INTRODUCTION TO THE DIGESTIVE SYSTEM

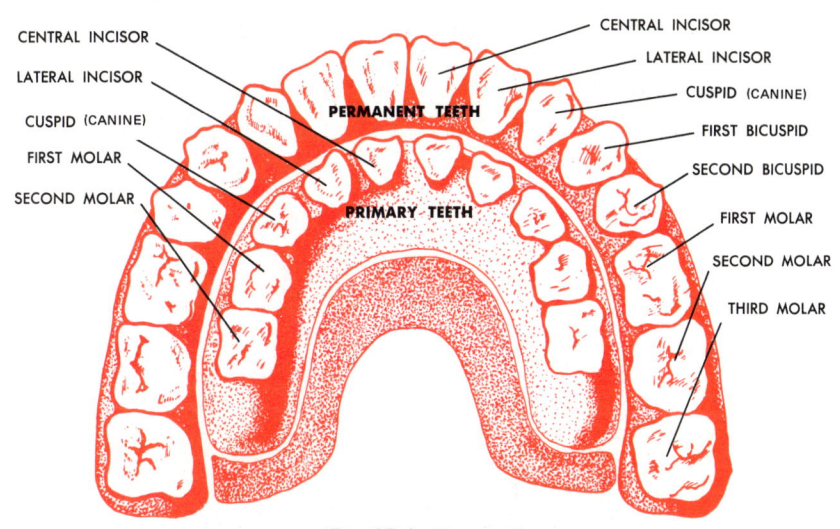

Fig. 25-3 The teeth

of each is particularly adapted to its purpose. Food must be chewed thoroughly so that the digestive juices and lubricating substances from the salivary glands may mix with the food particles; this begins the process of digestion.

SUGGESTED ACTIVITIES

- Discuss what measures are necessary for proper care of the teeth.
- Make a list of foods which are rich in calcium and phosphorus.
- Discuss other nutrients important in maintaining the health of the mouth.

REVIEW

A. Answer the questions, briefly.

1. What is the purpose of the salivary glands.

2. Name the four kinds of teeth and give the uses of each type.

DIGESTION OF FOOD

B. Match each of the terms in column I with its correct statement in column II.

Column I	Column II
_____ 1. papillae	a. substances that promote chemical reactions in living things
_____ 2. calcium and phorphorus	b. bleeding gums
_____ 3. digestion	c. a small soft structure suspended from the soft palate
_____ 4. the teeth	d. gums which protect the teeth
_____ 5. enzymes	e. tract consisting of the mouth, stomach and intestines
_____ 6. gingivae	f. minerals of which teeth are composed
_____ 7. accessory organs and structures of digestion	g. teeth, tongue, salivary glands, pancreas, liver, gallbladder and appendix
_____ 8. ptyalin	h. hardest structure in the body
_____ 9. uvula	i. projections on the surface of the tongue containing the taste buds
_____ 10. alimentary canal	j. the process of changing complex solid foods into soluble forms to be absorbed by cells
	k. number of teeth in the adult set
	l. the enzyme manufactured by the salivary glands

C. Label the teeth indicated on the diagram.

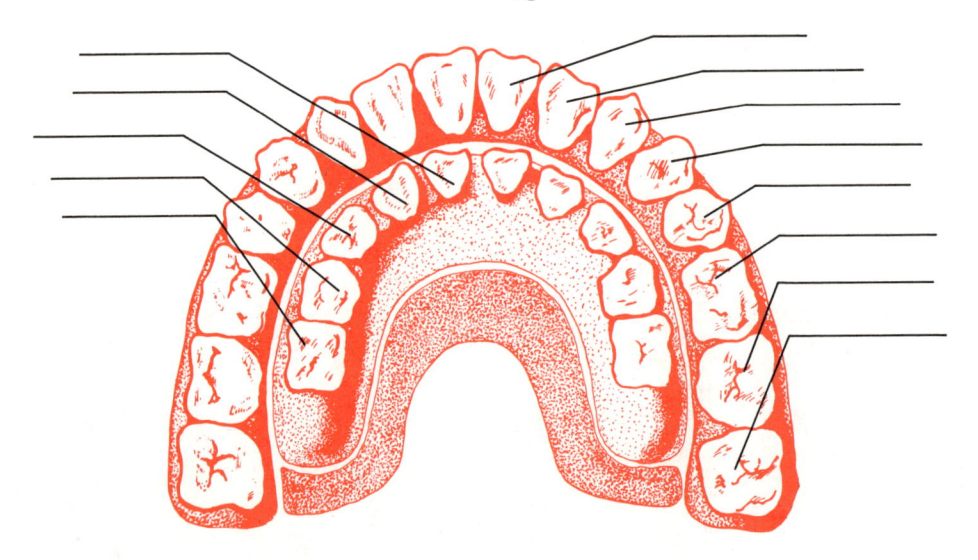

unit 26 digestion in the stomach

OBJECTIVES

Studying this unit should help the student:

- Describe functions of the pharynx, esophagus, and stomach.
- Explain the action of gastric juice.
- Describe the work of the various enzymes.

When food has been chewed and moistened with saliva in the mouth, it is swallowed by the action of the *pharynx*. It then passes through the esophagus, and enters the stomach by way of the cardiac sphincter, figure 26-1.

In addition to being a temporary storage place for food, the stomach has important digestive functions. It is a pear-shaped, elastic organ capable of stretching and shrinking to accommodate its contents. Stomach movements mix the contents of the organ and also move the food toward the intestine. The inner lining of the stomach contains thousands of tiny digestive glands called *gastric glands*. These manufacture gastric juice which contains primarily the enzyme *pepsin* and a small amount of hydrochloric acid. Pepsin is an enzyme that acts on proteins, breaking them into smaller components. Hydrochloric acid converts pepsinogen to its active form pepsin, dissolves some insoluble minerals, and destroys many bacteria that enter the stomach with food. It also regulates the action of the *pyloric valve* at the lower end of the stomach. Oversecretion of hydrochloric acid may irritate the stomach lining and lead to the development of a peptic ulcer.

The action of the gastric juice is helped by the churning action of the stomach walls. The semiliquid food which results is called *chyme*. When the chyme is ready to leave the stomach, a valve (*pyloric sphincter*) at the lower end of the stomach opens from time to time and allows the food to spurt on into the first part of the small intestine called the *duodenum*. The contraction and relaxation of smooth muscles in the walls of the alimentary tract move the food along the entire alimentary canal. This action, known as *peristalsis,* is an involuntary and automatic muscular action.

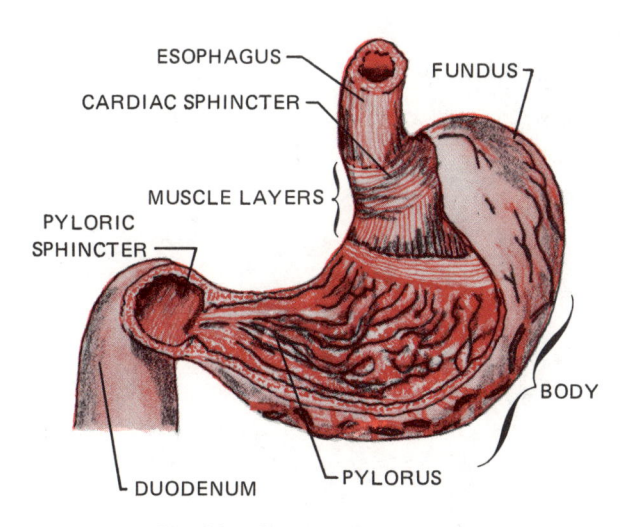

Fig. 26-1 Parts of the stomach

SUGGESTED ACTIVITIES

- Identify each organ of the digestive system on a wall chart or a torso model.

DIGESTION OF FOOD

- If laboratory facilities are available, place a small quantity of finely chopped, hardboiled eggwhite into each of four test tubes. To test tube #1, add 5 ml water; to test tube #2, add 5 ml 0.5% hydrochloric acid; to test tube #3, add a tiny amount of pepsin in 5 ml water; to test tube #4, add both a little pepsin and 5 ml of 0.5% hydrochloric acid. Place all the test tubes in an incubator overnight. Observe the results and complete the following table.

TEST TUBE	CONTENTS	OBSERVATION	CONCLUSION
1	White of egg plus water	No change	Water alone does not dissolve protein
2	White of egg plus 0.5% hydrochloric acid		
3	White of egg plus pepsin and water		
4	White of egg plus pepsin plus 0.5% hydrochloric acid		

- Add 5 ml of 0.5% hydrochloric acid to an equal volume of warm milk; note the reaction. Can this action be related to digestion?
- Explain why the stomach may change its shape before and after a meal.

REVIEW

A. Briefly answer the following questions.

1. List the organs of the alimentary canal and the accessory organs of digestion, discussed in the preceding unit.

   Alimentary Canal *Accessory Organs*

2. Explain the importance of hydrochloric acid in the stomach.

3. Name the main enzyme found in gastric juice and explain what it does.

DIGESTION IN THE STOMACH

B. Match each of the terms in column I with its correct statement in column II.

Column I	Column II
____ 1. chyme	a. semiliquid condition of food found in the stomach
____ 2. esophagus	b. acts upon protein in the stomach
____ 3. hydrochloric acid	c. oversecretion may cause peptic ulcer
____ 4. pepsin	d. involuntary muscle action of alimentary canal
____ 5. peristalsis	e. passageway to the stomach
____ 6. pharynx	f. storage place for food
____ 7. ptyalin	g. lower section of stomach
____ 8. pylorus	h. passage where swallowing action takes place
____ 9. stomach	i. enzyme which changes starch to sugar
____ 10. fundus	j. upper portion of stomach
	k. sphincter muscle at entrance to stomach

C. Label the parts of the stomach and the organs through which food enters and leaves the stomach.

unit 27 digestion in the small intestine

OBJECTIVES

Studying this unit should help the student:

- Describe how the small intestine prepares food for absorption.
- Describe the digestive function of the liver.
- Explain the digestive function of the gallbladder.

The small intestine is a coiled portion of the alimentary canal and is about twenty-five feet long and one inch in diameter. It contains thousands of small intestinal glands which produce intestinal juice. In addition to intestinal juice, bile from the liver and pancreatic juice from the pancreas are poured into the duodenum, the first part of the small intestine, figure 27-1.

Liver bile is needed for the digestion of fat. It breaks up the fat into small droplets upon which the digestive juices can act. Pancreatic juice contains enzymes that: (a) continue the digestion of protein started in the stomach, (b) act on starch, and (c) digest fat. The enzymes of the intestinal juice complete the digestion of proteins and carbohydrates. Thus the combined action of bile, pancreatic juice, and intestinal juice completes breaking down the food mass into substances which can be absorbed into the bloodstream.

The absorption is possible because the lining of the small intestine is not smooth. It is covered with millions of tiny projections called *villi*. Each microscopic villus contains a network of blood and lymph capillaries. The digested portion of the food passes through the villi into the bloodstream

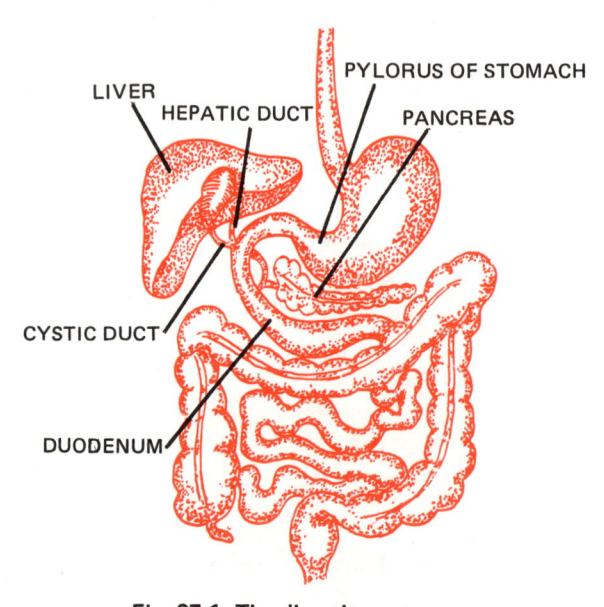

Fig. 27-1 The digestive process

DIGESTION IN THE SMALL INTESTINE

CARBOHYDRATES

STARCH → DOUBLE SUGARS (MALTOSE, LACTOSE, SUCROSE) → SIMPLE SUGAR (GLUCOSE)

FATS

FATS → EMULSIFIED FATS → FATTY ACIDS AND GLYCERIN

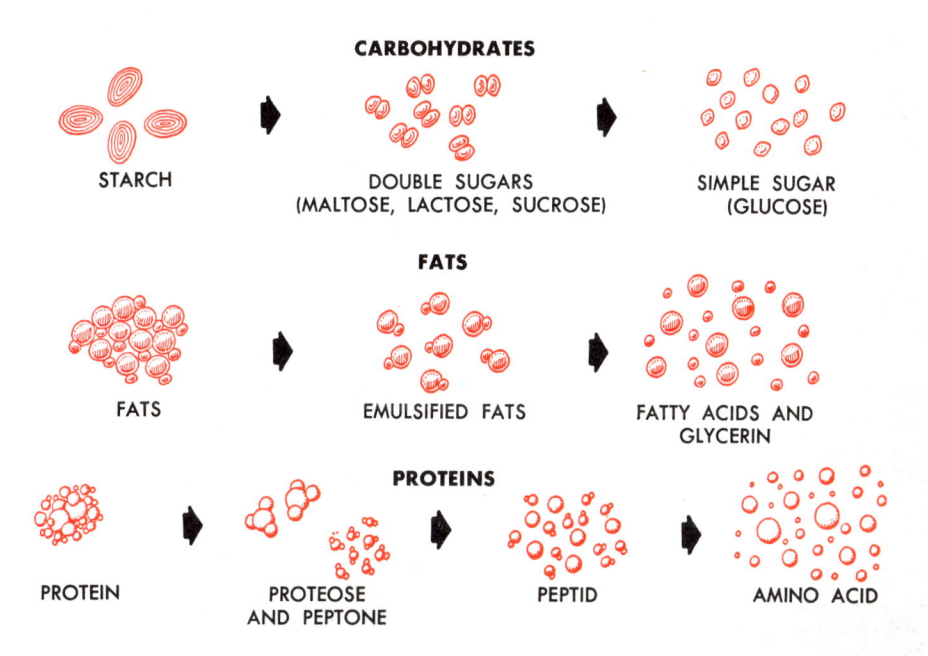

PROTEINS

PROTEIN → PROTEOSE AND PEPTONE → PEPTID → AMINO ACID

Fig. 27-2 Phases in digestion of starch, fat, and protein

and on to the body cells. The undigestible portion passes on into the large intestine.

THE LIVER AND GALLBLADDER

During the process of digestion, the liver, a large organ located just below the diaphragm on the right side, mainly acts on fat metabolism. It manufactures bile and passes it along to its storehouse, the gallbladder. When bile is needed for the digestion of fats, the gallbladder releases it through a duct into the duodenum of the small intestine.

In addition to manufacturing bile, the liver produces and stores glycogen (animal starch) from unused digested sugars. The liver also aids in removing certain waste products from the bloodstream, changing them into a form that can be excreted by the kidneys.

The bile contains mineral salts. If stored too long these salts may crystallize and form gallstones either in the gallbladder or in the ducts through which the bile passes. They may keep the bile from reaching the small intestine.

Figure 27-2 shows how starch, fat, and protein are broken into simple forms and made ready for absorption.

SUGGESTED ACTIVITIES

- Discuss reasons for giving hospital patients dilute glucose solution through their veins instead of a regular diet by mouth.

- Learn the meaning and use of the following terms:

villi	bile	absorption
lacteal	glycogen	intestinal juice
liver	pancreatic juice	gallbladder

DIGESTION OF FOOD

REVIEW

Match each of the terms in column I with its correct statement in column II.

Column I	Column II
_____ 1. small intestine	a. substances which contain enzymes capable of acting on the digestion of proteins, starch and fats
_____ 2. end result of protein digestion	b. tiny projections in the small intestines which greatly increase absorption area
_____ 3. villi	c. receives the undigested portion of food at the end of the alimentary canal
_____ 4. pancreatic juice	d. region into which bile from the liver and pancreatic juice are poured
_____ 5. enzymes of intestinal juice	e. amino acids
_____ 6. duodenum	f. fatty acids and glycerin
_____ 7. liver bile	g. sucrose
_____ 8. large intestine	h. is about 25 feet long and one inch wide
_____ 9. glucose	i. galactose
_____ 10. usable products of fat metabolism	j. digestive juice for fat metabolism
	k. substance which may result from the breakdown of starch
	l. enzymes which complete digestion of proteins and carbohydrates

unit 28 the large intestine

OBJECTIVES

Studying this unit should help the student:

- Locate the large intestine.
- Describe functions of the large intestine.
- List foods which aid the function of the colon.

The large intestine, or *colon,* is about five feet long and two inches in diameter. It extends from the small intestine to the anus, figure 28-1. The appendix is located a short distance from the point where the small intestine opens into the first part of the large intestine. The appendix is a narrow tube with only one opening. Its function is not known.

The large intestine is concerned with the storage and excretion of the waste products of digestion. It also aids in the regulation of the water balance of the body because its lining absorbs water. If the muscular activity of the large intestine is decreased, the waste products are not moved along. Constipation (abnormal infrequency of defecation) results.

Constipation may be avoided by exercise and by eating foods which contain bulk, such as cereals, fruits and vegetables. Regular bowel movements are of great importance to the general health.

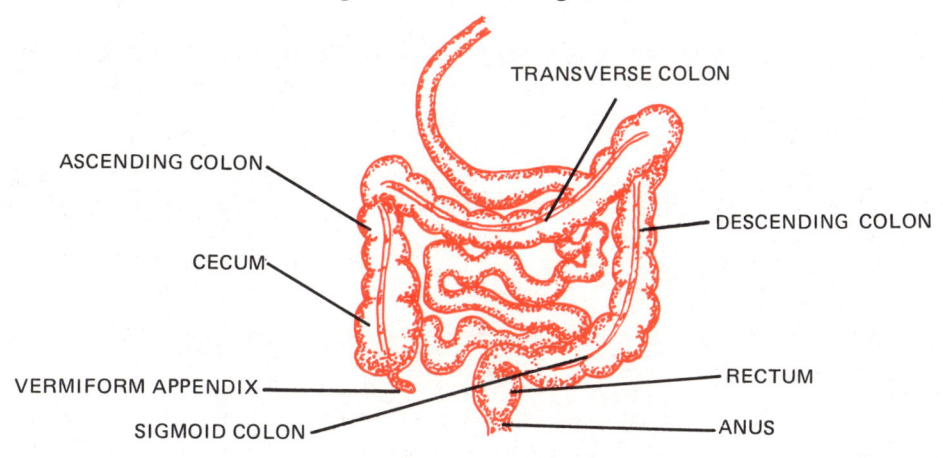

Fig. 28-1 The large intestine

SUGGESTED ACTIVITIES

- Discuss what effect waste products could have on the general health if they were not removed from the intestines.
- Discuss the special care needs of a person on bedrest, on the basis of this unit.
- Prepare a list of foods which are commonly recommended for a person who suffers from constipation.

DIGESTION OF FOOD

REVIEW

A. Briefly answer the following questions.

1. What causes constipation?

2. What is the function of the large intestine other than storage and elimination of wastes?

B. Read each statement carefully and determine if it is true or false. Encircle the letter *T* for true or *F* for false.

T   F   1. The large intestine is called the colon.

T   F   2. The large intestine is 20 feet long and 2 inches wide.

T   F   3. The cecum is located where the small intestine joins the large intestine.

T   F   4. The function of the appendix is unknown.

T   F   5. The large intestine stores and eliminates the waste products of digestion.

T   F   6. Regulation of water balance occurs in the large intestine because its lining absorbs water.

T   F   7. Constipation may be overcome by intensive and long periods of work and exercise.

T   F   8. Bulk foods such as cereals, fruits and vegetables may help avoid constipation.

T   F   9. The rectum is an extension of the descending colon.

T   F   10. The transverse colon lies between the ascending and the descending colon.

unit 29 disorders of the digestive system

OBJECTIVES

Studying this unit should help the student:
- Identify common disorders which interfere with digestion.
- Relate general treatment to these common disorders.

It is well to know the common disorders of the digestive system and the general treatment of them. Some of these are listed below.

Chronic constipation refers to the passage of dry, infrequent bowel movements. Treatment usually consists of eating proper foods, drinking plenty of liquids, getting enough exercise, setting regular bowel habits, and avoiding tension.

Diarrhea is the opposite of constipation. Frequent, liquid bowel movements may result from intestinal infection, poor diet, nervousness or from other more serious disorders.

Heartburn refers to a burning sensation in the esophagus from acid contents which have backed up from the stomach.

Stomatitis is an inflammation of the soft tissues of the mouth cavity. Pain and salivation may occur also.

Gastritis is an inflammation of the mucous membrane lining of the stomach. It may be caused by irritants such as highly spiced foods or some drugs.

Other less common disorders of the digestive system may require admittance to the hospital. *Peptic ulcers* are lesions which occur in either the stomach or intestinal walls. These ulcers are thought to result from increased secretion of hydrochloric acid in times of stress. Rest, reduction of stress and change of diet often help, although surgery may be needed.

Hiatal hernia, or rupture occurs when the stomach protrudes above the diaphragm through the esophagus opening. Changes in the diet may relieve the heartburn; surgery is not usually required.

Pyloric stenosis is a condition in which the pyloric sphincter narrows. It is often found in infants. Projectile vomiting may result; surgery is often needed.

Gastroenteritis is the inflammation of the mucous membrane lining of the stomach and intestinal tract. This is a common disorder of infants which may lead to severe diarrhea and dehydration.

Infectious hepatitis is a viral infection of the liver, often spread through contaminated water or food. A common symptom is jaundice, a yellowish tinge to the skin caused by bile in the bloodstream.

Cirrhosis is a chronic progressive disease of the liver. It is commonly caused by lack of proper food.

Peritonitis is a condition in which the lining of the abdominal cavity is inflamed. Vomiting and pain are symptoms of this condition.

Gallstones are collections of crystallized salts which form in the gallbladder. Pain and digestive disorders may occur. Surgery may be necessary if a change in diet along with other treatments does not bring relief.

Cholecystitis is the inflammation of the lining of the gallbladder sometimes causing blockage of the cystic duct.

Carcinoma or cancer may occur in any part of the digestive tract. Surgery, radiation, or chemotherapy are prescribed.

DIGESTION OF FOOD

SUGGESTED ACTIVITIES
- Discuss problems which result from relieving symptoms such as heartburn and diarrhea with over-the-counter drugs instead of seeking medical advice.

REVIEW
Match each of the terms in column I with its correct statement in column II.

Column I	Column II
_____ 1. cirrhosis	a. frequent liquid bowel movements
_____ 2. gastroenteritis	b. chronic liver disease
_____ 3. peptic ulcers	c. protrusion of the stomach into the esophagus opening
_____ 4. hiatal hernia	d. viral infection of the liver
_____ 5. heartburn	e. inflammation of the abdominal cavity
_____ 6. diarrhea	f. obstruction of the hepatic duct
_____ 7. cholecystitis	g. inflammation of the stomach and intestinal lining
_____ 8. infectious hepatitis	h. inflammation of the gallbladder lining
_____ 9. pyloric stenosis	i. narrowing of sphincter in the stomach
_____ 10. peritonitis	j. cardiospasm
	k. lesions which result from acid secretion in times of stress
	l. common symptom characterized by a burning sensation

section 6 digestion of food
SELF-EVALUATION

A. Label the following diagram of the digestive system.

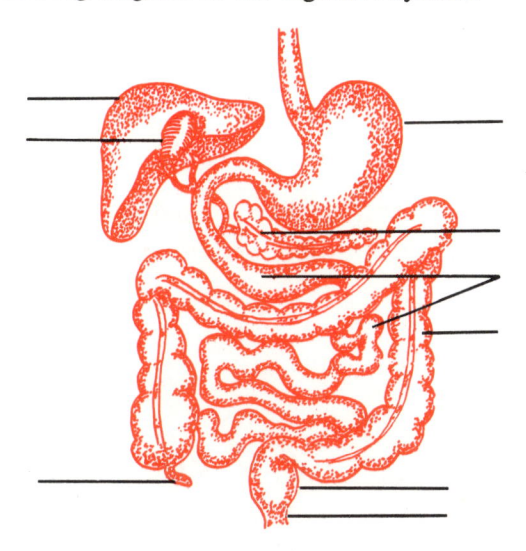

B. Complete the following statements.

1. Substances which act chemically upon foods to change them to simpler soluble forms are called _____ .

2. Teeth used for biting or cutting food are _____ ; those used for grinding are _____ ; and those used for tearing are _____ .

3. Digested food enters the bloodstream by passing through the _____ in the _____ intestine.

4. The main functions of the large intestine are _____ and _____ .

5. The three juices which act upon food in the small intestine are _____ , _____ , and _____ .

C. Select the item which best completes the statement and encircle the letter before it.

1. One part of the small intestine is the
   a. rectum
   b. duodenum
   c. pancreas
   d. appendix

2. The appendix is attached to the
   a. duodenum
   b. rectum
   c. cecum
   d. pylorus

93

SELF-EVALUATION

3. Bile is secreted by the
   - a. pancreas
   - b. gallbladder
   - c. stomach
   - d. liver

4. A substance which requires further digestion to break it down is
   - a. fatty acid
   - b. amino acid
   - c. protein
   - d. glycerol

5. The salivary glands are situated
   - a. in the small intestine
   - b. in the pancreas
   - c. in the mouth
   - d. in the stomach

D. How does the blood differ in its composition before it enters the capillaries of the small intestine and after it leaves?

E. Explain the function of the liver and gallbladder in digestion.

SECTION 7

ELIMINATION OF WASTE MATERIALS

unit 30 introduction to the excretory system

OBJECTIVES

Studying this unit should help the student:

- Explain the function of the excretory organs.
- List the parts of the body involved in elimination.
- Relate the type of waste to the channel of excretion.

Food is utilized through the process of digestion, absorption, and metabolism. These steps of the process separate those substances which can be digested from those which cannot. The blood and lymph transport the products of digestion to the tissues where they are needed. After the cells of the tissues have used the food and oxygen needed for growth and repair, the waste products formed are taken away by the blood and excreted from the body. If they were left to accumulate in the body they would act as poisons. The excretory organs eliminate the metabolic wastes and undigested food residue.

The channels through which excretion takes place include the kidneys, the skin, the intestines, and the lungs. The lungs, generally considered part of the respiratory system, serve an excretory function in that they give off carbon dioxide and water vapor in exhalation. The urinary system functions largely as an excretory agent of nitrogenous wastes, salts, and water, while the skin includes excretion of the dissolved wastes present in perspiration, mostly dissolved salts. The indigestible residue, water and bacteria are excreted by the intestines. These body excretions are summarized in figure 30-1.

ORGAN	PRODUCT OF EXCRETION	PROCESS OF ELIMINATION
Lungs	carbon dioxide	exhalation
Kidneys	nitrogenous wastes and salts dissolved in water	urination
Skin	dissolved salts	perspiration
Intestines	solid wastes	defecation

Fig. 30-1 The excretion of waste products

SUGGESTED ACTIVITIES

- Discuss what may happen to a person if excretion does not continue properly.
- Discuss which part of the excretory system performs the most important function.

ELIMINATION OF WASTE MATERIALS

- Define the terms: dysuria, excretion, incontinence, micturition, retention.

REVIEW

Briefly answer the following questions.

1. What is the function of the excretory organs?

2. What organs are involved in excretion?

3. How are waste products transported to the organs of excretion?

4. Name the excreted waste products of the body.

5. Explain the excretory function of the lungs.

unit 31 urinary system

OBJECTIVES

Studying this unit should help the student:
- List the organs which make up the urinary system.
- Describe how the kidneys excrete wastes from the body.
- Explain how the kidneys regulate the water balance.

The urinary system performs the main part of the excretory function in the body. It consists of the kidneys, ureters, bladder, and urethra, figure 31-1. The kidneys are two bean-shaped organs which lie behind the abdominal cavity, separated from it by the peritoneum, one on each side of the spinal column.

The kidneys serve as filters of the bloodstream. The renal artery which branches directly from the aorta supplies blood to the kidneys. The renal vein returns blood to the inferior vena cava. As the blood passes through the kidneys, nitrogenous wastes, excess salts, and excess water are removed from it by millions of tiny filters called *nephrons*.

A nephron consists of the glomerulus, Bowman's capsule, loop of Henle, and collecting tubules. As the blood goes through the coiled, knoblike mass of capillaries called the *glomerulus,* waste substances from it are absorbed by the *Bowman's capsule.* Then these substances pass through the *tubules* where much of the water and certain minerals are reabsorbed into the blood. Only the nitrogenous wastes, excess water, and excess mineral salts (urine) continue through the tubules, which drain into the pelvis of the kidney *(renal pelvis).*

The urine passes from the kidneys to the bladder through tubes called *ureters.*

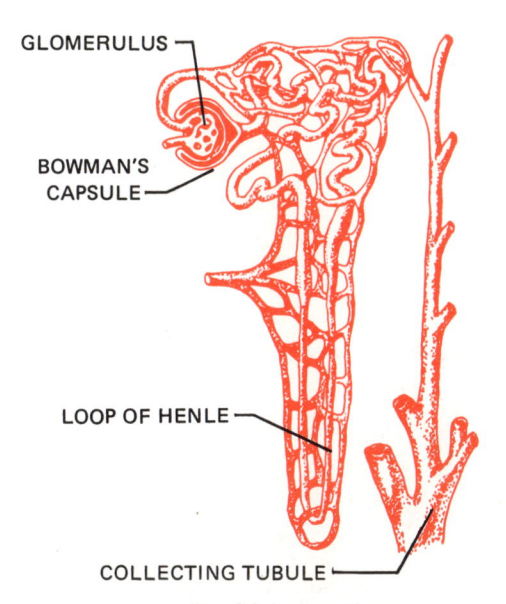

Fig. 31-1 A nephron

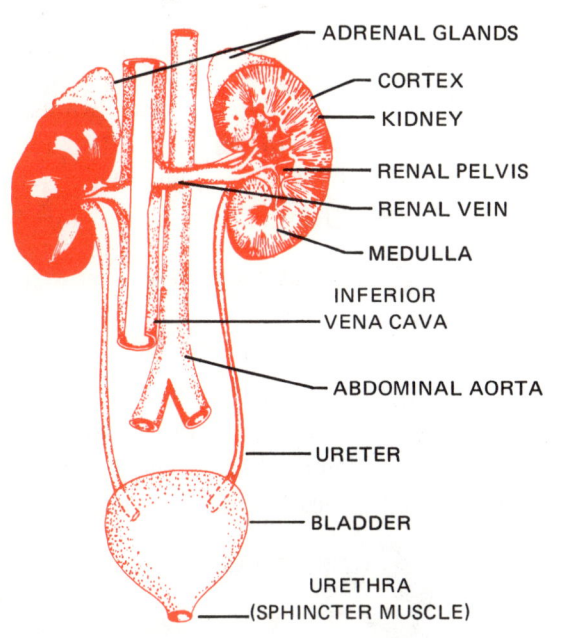

Fig. 31-2 The urinary system

ELIMINATION OF WASTE MATERIALS

The *bladder,* a hollow muscular organ, acts like a reservoir. It stores the urine until about one pint is accumulated. The bladder then becomes uncomfortable and must be emptied or voided. Voiding takes place by muscular contractions of the bladder which are involuntary, although they can be controlled to some extent through the nervous system. Contraction of the bladder muscles forces the urine through a narrow canal, the *urethra,* which extends to the outside opening.

In addition to removing metabolic wastes from the body, the kidneys also have the important function of regulating water balance in the body. Each day the blood carries over 150 quarts of water, waste, and dissolved food substances through the kidneys, depending on the amount of fluid intake of the person. All but about 3 pints returns purified, to the bloodstream. The balance is excreted as urine, carrying waste products filtered from the blood. The reabsorption of water is largely due to control by the ADH (antidiuretic) hormone from the posterior lobe of the pituitary gland.

The kidneys have the potential to work harder than they actually do. Under ordinary circumstances only a portion of the glomeruli are used. Should one kidney not function, or have to be removed, more glomeruli and tubules open up in the second kidney to assume the work of the nonfunctioning or missing kidney.

SUGGESTED ACTIVITIES

- If laboratory facilities are available, obtain and examine several specimens of fresh, normal urine.

    a. What is the color of the specimen?

    b. Is it clear or cloudy?

    c. Is the urine acid, alkaline or neutral?

    To test, dip blue litmus paper into the urine. If acid is present, it will turn red. Dip red litmus paper in. If urine is alkaline, it will turn paper blue. If neither paper changes color, the urine is neutral.

    d. What is the specific gravity of a specimen? To test, use a urinometer.

    e. Is albumin present?

    To test for albumin, place 10 milliliters of urine in a test tube. Add 3 drops of dilute acetic acid (2%). Hold the tube at the bottom and apply heat to the upper level of urine. If a cloud appears in the heated portion, albumin is present.

    f. Is sugar (glucose) present?

    To test for sugar, place 10 drops of urine in a Pyrex test tube. Add 5 milliliters of Benedict's solution. Mix thoroughly by shaking gently.

Place in a water bath and boil for 3 minutes. As soon as the bubbling stops, interpret the test results as follows:

COLOR	INDICATION
No change in color	0 absent
Green	± trace of sugar
Greenish-yellow	+ one plus
Yellow	++ two plus
Brown or brick red	+++ three plus

- Using Acetest reagent tablets, examine the urine for acetone. Have the results and your interpretation checked by the instructor.

  Place the reagent tablet on a clean white sheet of paper. Place a drop of urine on the tablet. In 30 seconds, compare the resulting color with the color chart enclosed with the tablets. Record the result on the chart.

- Using Clinitest tablets and/or Clinistix reagent strips, test for sugar. Have the results and your interpretation checked by the instructor.

  Clinitest tablets: Place 5 drops of urine and 10 drops of water in a test tube. Add the Clinitest tablet. Observe the reaction. Then shake the test tube and compare the color of the solution with the color scale enclosed with the tablets. Record the result.

  Clinistix reagent strips: Dip the test end of the Clinistix in the urine and remove it. (Avoid contact with fingers or other objects because misleading results may occur.) If the moistened end turns blue, the result is *positive*. When sugar is present, the blue color will appear in less than one minute. Record the result.

- Using Bumintest reagent solution and/or Allritest tablets, test the urine for albumin. Have the results and your interpretation checked by the instructor.

  Bumintest: Dissolve 4 Bumintest reagent tablets in 30 milliliters of water in a test tube. (This makes a 5% solution.) In another test tube, mix equal parts of urine and Bumintest solution and shake the tube gently. The amount of albumin is estimated by the degree of cloudiness (turbidity). Record the result.

  Allritest tablets: Place Allritest tablet on clean paper. Put one drop of urine on the tablet. When the urine has been absorbed, add 2 drops of water and allow the water to be absorbed before reading. Compare the color of the top of the tablet with the color photograph enclosed with each package of tablets. Record the result.

ELIMINATION OF WASTE MATERIALS

REVIEW

A. Match each of the terms in column I with its correct statement in column II.

Column I	Column II
___ 1. nephron	a. tubes which connect the kidney with the bladder
___ 2. glomerulus	b. mass of capillaries
___ 3. bladder	c. structure which absorbs wastes from the capillary mass
___ 4. urethra	d. one of millions of tiny filtering units
___ 5. ureter	e. returns blood to the inferior vena cava
___ 6. ADH	f. hormone which regulates water reabsorption
___ 7. tubules	g. contraction of bladder muscles
___ 8. Bowman's capsule	h. canal which opens to the outside of the body
___ 9. kidney	i. primarily acts as a reservoir
___ 10. renal vein	j. allow urine to drain into the renal pelvis
	k. bean-shaped organ

B. Answer the following questions briefly.

1. How do the kidneys function in excretion?

2. Of what value is it to the physician to have an exact measure of the patient's intake and output?

3. What is another important function of the kidneys besides the elimination of waste?

unit 32 the skin

OBJECTIVES

Studying this unit should help the student:
- Describe the functions of the skin.
- Explain how the skin serves as a channel of excretion.
- Describe the action of the sweat glands.

Although the skin is the most visible of the channels of excretion, excretion is only one of its several functions. Skin is first thought of as a covering for the underlying structures to protect them from injury and germ invasion. The skin also helps regulate body temperature by controlling the amount of heat loss. Evaporation of water, in the form of perspiration, from the skin helps rid the body of excess heat.

The skin is sometimes called the integument or the integumentary system. It serves as a sense organ in which sensory nerve endings detect heat, cold, pain, touch, and pressure. It also absorbs certain drugs and other chemical substances. The administration of drugs through the skin is called *inunction.* This absorptive quality can be harmful if insecticides, gas, and lead salts enter the body through the skin.

Actual excretion is a minor function of the skin; certain wastes dissolved in perspiration are removed. Perspiration is 99 percent water with only small quantities of salt and organic materials (waste products). Sweat, or *sudoriferous,* glands are distributed over the entire skin surface; they are present in large numbers under the arms, on the palms of the hands, soles of the feet, and on the forehead.

Sweat glands are tubular, with a coiled base and a tubelike duct which extends to form a pore in the skin, figure 32-1. Perspiration is excreted through the pores. Under the control of the nervous system, these glands may be activated by several factors, heat, pain, fever, and nervousness.

The amount of water lost through the skin is almost 500 ml a day. However, this varies according to the type of exercise and the environmental temperature. In profuse sweating a great deal of sodium chloride (salt) may be lost; it is vital to replace this.

The skin is protected by a thick oily substance known as *sebum* which is secreted by the *sebaceous glands.* Sebum lubricates the skin, keeping it soft and pliable.

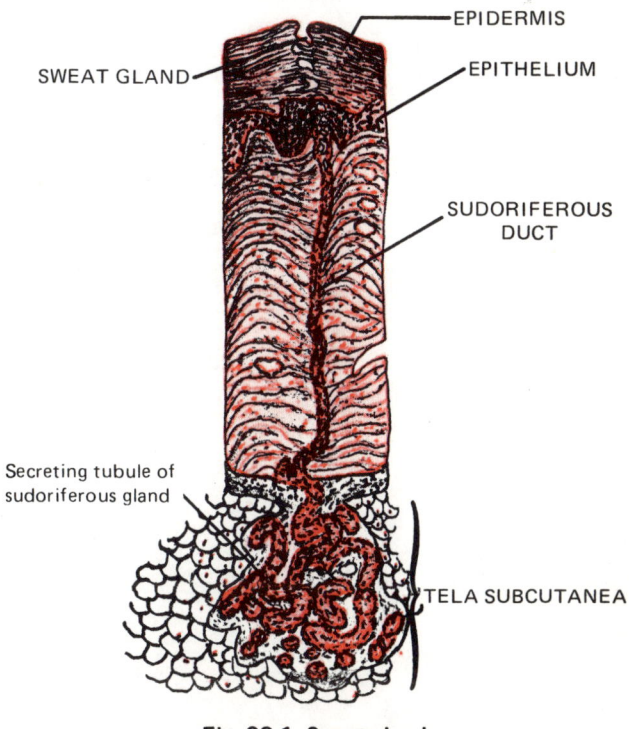

Fig. 32-1 Sweat gland

ELIMINATION OF WASTE MATERIALS

SUGGESTED ACTIVITIES

- Pour an equal amount of cold water in two beakers. Cover the water in one beaker with olive oil to prevent evaporation of water. Place them in a container of boiling water.

    a. In which beaker did the temperature rise more rapidly? Why?

    b. What conclusion may be drawn from this experiment in discussing heat loss and the regulation of body temperature in human beings?

- Take your temperature and record the result.

    a. By mouth

    b. By axilla

    c. By mouth, immediately after a cold drink

    d. By mouth, immediately after a hot drink

    Compare the various temperatures. Discuss the reasons for the differences.

REVIEW

1. In what way does the skin act as an organ of excretion?

2. What factors stimulate the sweat glands into activity?

3. Under normal condition, approximately how much water is lost daily through the skin?

4. What is perspiration?

5. What are the sudoriferous glands?

unit 33 disorders of the excretory system

OBJECTIVES

Studying this unit should help the student:
- List some disorders of the urinary tract.
- Recognize some common skin disorders.
- Describe the more common excretory system disorders.

The normal functioning of the excretory system has been described. There are also malfunctions and disorders of the urinary tract and the skin with which the nurse should be familiar.

DISORDERS OF THE URINARY TRACT

Acute kidney failure may be sudden in onset. Causes may be nephritis, shock, injury, bleeding, sudden heart failure or poisoning. A common symptom is the absence of urine, which is termed *anuria.* An artificial kidney machine may be used to remove wastes normally excreted by a healthy kidney; this process is called hemodialysis.

Cystitis is inflammation of the mucous membrane lining the bladder. It is usually caused by an infection. Frequent and painful urination results.

Pyelitis is inflammation of the pelvis of the kidney, usually due to an infection.

Pyelonephritis is the spread of bacteria into the kidney tissue. This produces the pus which is present in the urine.

Nephritis is inflammation of the kidney which causes damage to the tissue. The result of this condition is that the kidneys are unable to carry on the task of excretion in an efficient way. (Also called Bright's disease.) *Acute nephritis* usually occurs in children and young people. It may be a complication of a communicable disease, especially scarlet fever. The streptococcus organism may be the cause. *Chronic nephritis* is a kidney condition which develops gradually in older people. Usually high blood pressure is also present. Hardening of the renal blood vessels may be the cause, or the glomeruli and tubules may have been destroyed over an extended period of time.

Kidney stones are deposits of crystalline products, usually calcium, in the kidneys. The stones may be found in the urinary tract anywhere from the kidney to the bladder. There are various causes for the formation of these stones (calculi): extended immobility, dehydration, renal infection, or hyperparathyroidism may be contributing factors.

Tuberculosis of the kidney is a destructive kidney disease caused by the tubercle bacillus.

Uremia is an illness caused when the kidneys are unable to eliminate the waste products of metabolism. A continued piling up of waste products in the blood can lead to coma and death. Hemodialysis helps to remove wastes in many uremia cases.

DISORDERS OF THE SKIN

Acne vulgaris is a chronic disorder of the sebaceous glands. It occurs most often during adolescence and is marked by pimples, blackheads, cysts, and scarring.

Athlete's foot is a contagious fungus infection of the feet, usually contracted in public baths and/or showers.

Eczema is an allergic condition caused by diet, clothing, creams or soaps. The skin becomes dry, itchy, and scaly.

Gangrene is death of the tissue cells, caused by the blockage of blood supply to an area.

Impetigo contagiosa is a contagious skin disease seen in babies and young children, caused by the staphylococcus or streptococcus organism.

Pruritus is itching which may indicate a skin disease. Diabetes mellitus, liver ailments, and thyroid disorders may cause pruritus.

Psoriasis is a chronic disease characterized by reddened areas covered by silvery scales. It does not appear on the face.

Ringworm is a contagious fungus infection marked by red circular patches with crusts, which occur on skin and/or in the hair.

Scabies is also called "seven-year itch". It is caused by tiny parasites which get under the outer layer of the skin. Specific ointments, baths, and change of clothing are prescribed.

Urticaria, or hives, is a sudden appearance of edematous, raised, pink areas which itch and burn. It is usually caused by an allergy.

Furuncles are boils which are usually the result of staphylococcus infections in the hair follicle.

Carbuncles are deep abscesses. Treatment may require incision, drainage, and use of antibiotics.

Shingles (herpes zoster) is a skin eruption thought to be due to a virus infection of the nerve endings. It it commonly seen on the chest or abdomen.

SUGGESTED ACTIVITIES

- Explain the value of urinalysis in diagnosing cystitis or pyelitis.

- Inspect the skin on various areas of the body. Look for the skin pores, skin irritations, pimples, scaliness, and dryness. Discuss methods of proper cleansing of the skin; nutrition to improve the skin; the effect of sunshine on the skin; and sensitivity of skin to cosmetics.

- Visit a hemodialysis center. Ask for permission to read a patient's history. Write a short report to present to the class.

REVIEW

1. What may cause cystitis or pyelitis?

DISORDERS OF THE EXCRETORY SYSTEM

2. What is a frequent result of nephritis?

3. How are kidney stones formed?

4. What is the value of hemodialysis to the patients whose kidneys are unable to remove waste products from the blood?

section 7 elimination of waste materials
SELF-EVALUATION

A. Match each description in column I with the correct term in column II.

Column I	Column II
_____ 1. waste product eliminated through lungs	a. anuria
_____ 2. blood filter	b. calculi
_____ 3. stones in the kidney	c. carbon dioxide
_____ 4. water and nitrogenous wastes	d. cystitis
_____ 5. inflammation of the mucous membrane lining the bladder	e. kidneys
_____ 6. kidney malfunction preventing elimination of metabolic wastes	f. nephritis
_____ 7. helps regulate body temperature	g. perspiration
_____ 8. urinary duct	h. uremia
	i. ureter
	j. urine
	k. urinometer

B. Label the parts of the urinary system on the diagram.

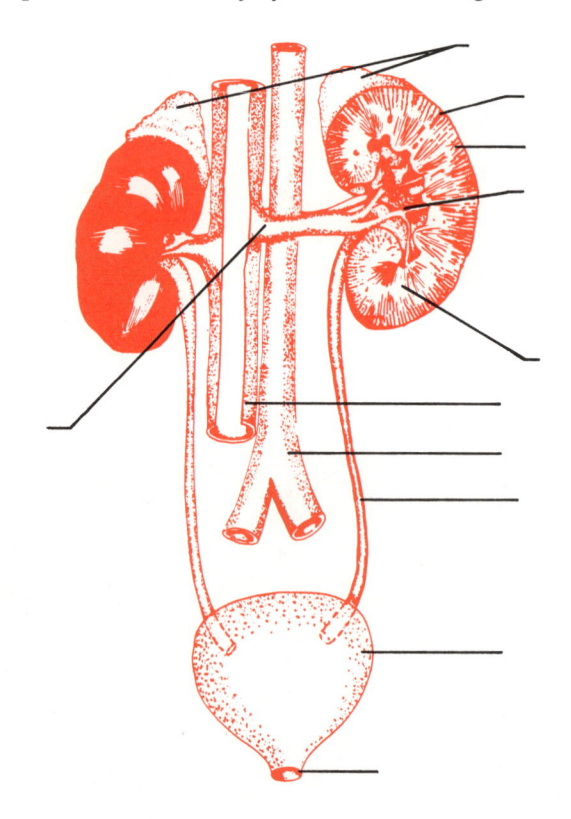

SELF-EVALUATION

C. Match each term in column I with its correct description in column II.

Column I	Column II
____ 1. acne vulgaris	a. a skin disorder of adolescence, marked by pimples, blackheads, cysts, scars
____ 2. pruritus	b. chronic skin disease characterized by red areas covered with silvery patches
____ 3. eczema	
____ 4. impetigo contagiosa	c. 7-year itch caused by tiny parasites which bore under outer layer of skin
____ 5. psoriasis	d. allergic condition sometimes due to diet, soap, or creams
____ 6. ringworm	
____ 7. shingles	e. a contagious skin disease of babies or young children caused by the streptococcus or staphylococcus organism
____ 8. urticaria	
____ 9. scabies	f. itching of the skin, may be due to diabetes mellitus, liver, or thyroid disorders
____ 10. carbuncle	
	g. contagious fungus infection with red circular crusty patches on skin or in hair
	h. sudden-appearing, raised pink areas which itch and sting; commonly called hives
	i. herpes zoster, skin eruption due to virus infection of nerve endings
	j. deep abscess requiring incision, drainage, and antibiotics
	k. death of tissue cells

SECTION 8

HUMAN REPRODUCTION

unit 34 introduction to the reproductive system

OBJECTIVES

Studying this unit should help the student:

- Contrast reproduction of simple cells and more complex forms of life.
- Explain the process of fertilization.
- Describe how physical traits are determined.

Regardless of the kind or complexity of the organism, the first step of reproduction occurs at the cellular level with the process of cell division. It may lead directly to the formation of specialized sex cells known as *gametes*. In gamete formation, a different type of cell division occurs than that discussed before in relation to tissue growth or replacement. It is called *meiosis*, or reproduction division, and occurs only in the reproductive organs. Meiosis produces cells that have only one of each pair of the 46 chromosomes

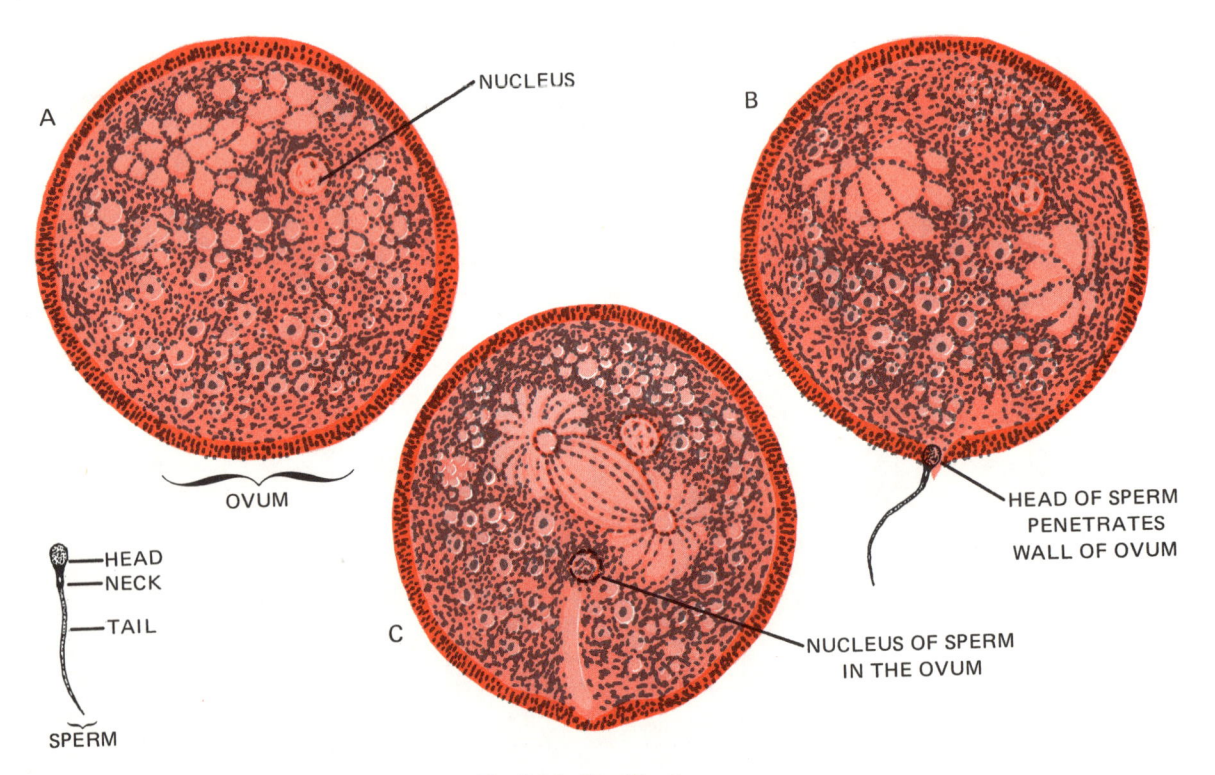

Fig. 34-1 Fertilization

INTRODUCTION TO THE REPRODUCTIVE SYSTEM

(heredity carriers) that occur in the body cells.

All complex organisms start from a fusion of two gametes: one from the male, called the sperm (spermatozoan), and one from the female, called the egg (ovum). During sexual intercourse sperms are transferred from the male to the female reproductive organs. They swim to the ovum that has been released from the ovary, figure 34-1. Only one sperm fuses with the ovum (a process called fertilization) to form the *zygote,* the beginning cell of a new individual.

Fertilization restores the full complement of 46 chromosomes possessed by every human cell, each parent contributing one chromosome to each of the 23 pairs. A substance called deoxyribonucleic acid (DNA) is found in the chromosomes; it contains the genetic code that is replicated and passed on to each cell as the zygote divides and redivides to form the embryo. This process continues until the new individual is completely formed.

All of the inherited traits possessed by the offspring are established at the time of fertilization. This is a point to remember when working with parents. A young mother may hope that her coming baby will be a girl with curly hair, or a father may insist that he wants a son. The nurse may assure them that the sex of the child and physical characteristics, such as eye color and curly hair, were determined at the time of fertilization. Concentration on these points cannot change the outcome.

SUGGESTED ACTIVITIES

- Discuss recessive and dominant heredity traits. Give examples of each.
- Find out if mental illness, tuberculosis, or cancer is inherited. Read about one of these and submit a paper on it.
- Learn the meaning and proper use of the following terms:

conception	primipara	semen
gametes	multipara	genetics
gonads	congenital	peritoneal cavity
inherited	pelvic cavity	zygote

REVIEW

1. When is the sex of the offspring determined?

2. What are chromosomes?

3. How many chromosomes are present in each body cell of the newborn child?

4. Where do these chromosomes come from?

5. What part does DNA play in hereditary characteristics?

unit 35 the organs of reproduction

OBJECTIVES

Studying this unit should help the student:

- Identify the organs of the female reproductive system.
- Identify the organs of the male reproductive system.
- Describe the functions of the reproductive organs.

The function of the reproductive system is to provide continuity of the species. In the human, the female reproductive system is composed of two ovaries, two fallopian tubes, the uterus, and the vagina. The male reproductive system is made up of two testes, seminal ducts, glands, and the penis. The principal male organs are located outside the body in contrast to the female organs which are largely located within the body.

FEMALE REPRODUCTIVE SYSTEM

The ovaries produce the female gametes, or *ova*. Ovaries contain thousands of microscopic sacs called *graafian follicles* in varying stages of development. Inside each follicle, an ovum develops. Usually only one follicle matures every twenty-eight days throughout the reproductive period of a woman, but occasionally two or more follicles may mature thus liberating more than one egg. (Gamete production ceases with menopause). As the follicle enlarges, it migrates to the outside surface of the ovary, breaks open, and the ovum is released from the ovary. This process is called *ovulation* and occurs about two weeks before the menstrual period begins; however, this time may vary somewhat with different individuals. After ovulation, the ovum makes its way down the *fallopian tubes*. Fertilization can take place within a few days after ovulation only, an important fact

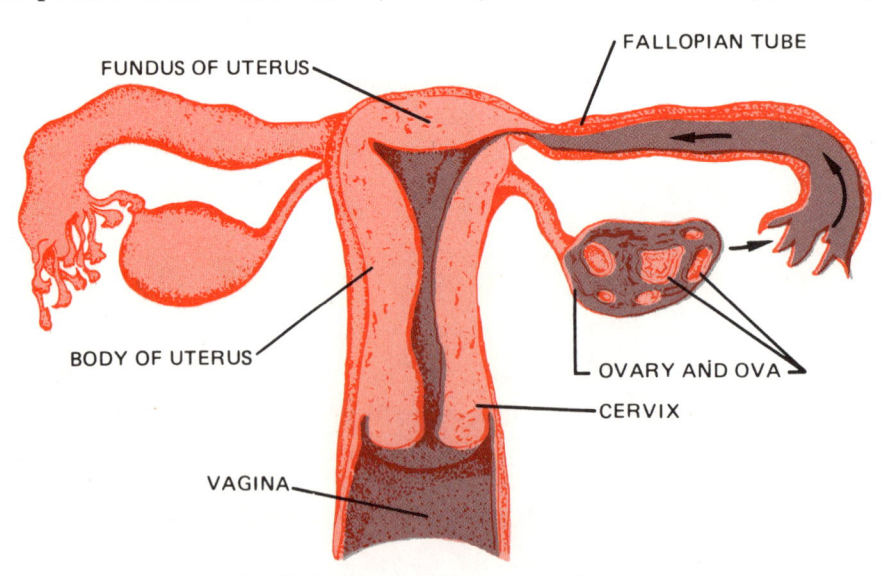

Fig. 35-1 Uterus, tubes, and ovaries

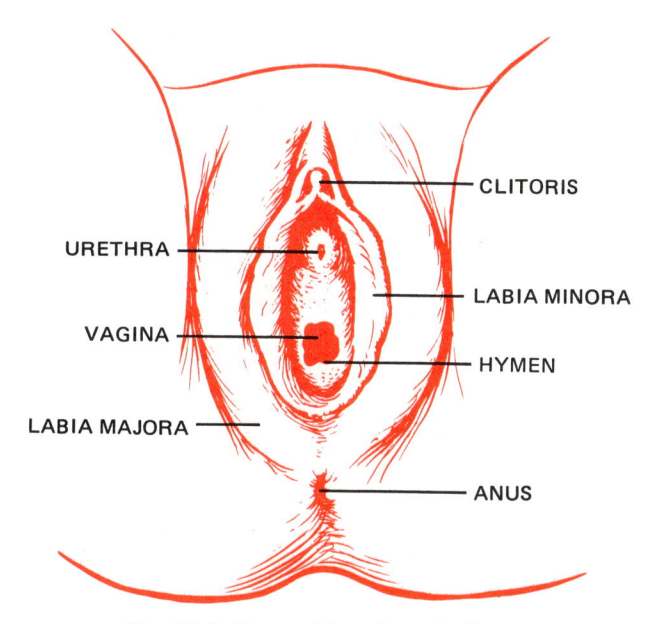

Fig. 35-2 External female genitalia.

in family planning. If fertilization does not take place, the ovum passes out of the body during menstruation.

The development of the follicle and release of the ovum is under the influence of two hormones produced in the anterior lobe of the pituitary gland, the follicle-stimulating hormone (FSH) and the luteinizing hormone (LH). FSH also promotes the secretion of *estrogen* by the ovary. Estrogen promotes the rapid growth of the lining of the uterus (the endometrium) in preparation for possible implantation of a fertilized ovum.

Following ovulation, the ruptured follicle enlarges, takes on a yellow fatty substance and becomes the *corpus luteum* ("yellow body"). The corpus luteum secretes another ovarian hormone, *progesterone,* which functions to maintain the growth of the uterine lining. If the egg is not fertilized, the corpus luteum degenerates, progesterone production stops, and the thickened glandular endometrium sloughs off. The tiny blood vessels that supply the endometrium are ruptured producing the blood flow of menstruation. Following menstruation the endometrium heals and starts thickening again marking the beginning of the new cycle.

The fallopian tubes, about four inches long and not attached to the ovaries, serve as conveyors or ducts for the ovum on its way to the *uterus,* a muscular organ about the size of a pear. The wide upper part is called the *fundus;* the lower part, the *cervix;* and the middle portion is the *body*.

After fertilization, which usually occurs in the fallopian tubes, the zygote becomes embedded in the uterus where it undergoes development into the *embryo* and later the *fetus.*

During development of the embryo-fetus, the uterus gradually rises until the top part is high in the abdominal cavity, pushing on the diaphragm. This may cause some difficulty in breathing in the late stages of pregnancy.

The *vagina* is the short canal which extends from the cervix of the uterus to the vulva. It is muscular tissue which receives the sperm during sexual intercourse and stretches during childbirth.

The breasts are accessory organs to the female reproductive system. They are composed

HUMAN REPRODUCTION

of many lobes arranged in a circular formation. Clusters of secreting cells surrround tiny ducts. One duct, only, goes from each lobe to an opening in the nipple. The *areola,* the colored area which surrounds the nipple, changes to a brownish color during pregnancy. *Prolactin* from the anterior lobe of the pituitary gland stimulates the mammary glands to secrete milk following the birth of a baby.

MALE REPRODUCTIVE SYSTEM

The male reproductive organs consist of the testes, a system of ducts for transporting the sperm from the testes, several glandular structures, and the penis.

The *scrotum* is an external sac which contains two ovoid bodies called testes. They are the primary sex organs of the male and produce male gametes, the *sperm,* and the male sex hormone, *testosterone.* This hormone influences the development of the secondary sexual characteristics of the male — deep voice; facial, pubic, and axillary hair; the typical male shape of wide shoulders and narrow hips — and is also necessary for the normal development of the secondary male reproductive organs: *epididymis, vas deferens, seminal vesicles, prostate gland,* and *penis,* figure 35-3.

The *gonadotrophic* hormone from the anterior lobe of the pituitary gland stimulates the testes to activity. Certain interstitial cells of the testes produce the testosterone. Sperms are formed in the *seminiferous tubules* of the testes. The process of sperm

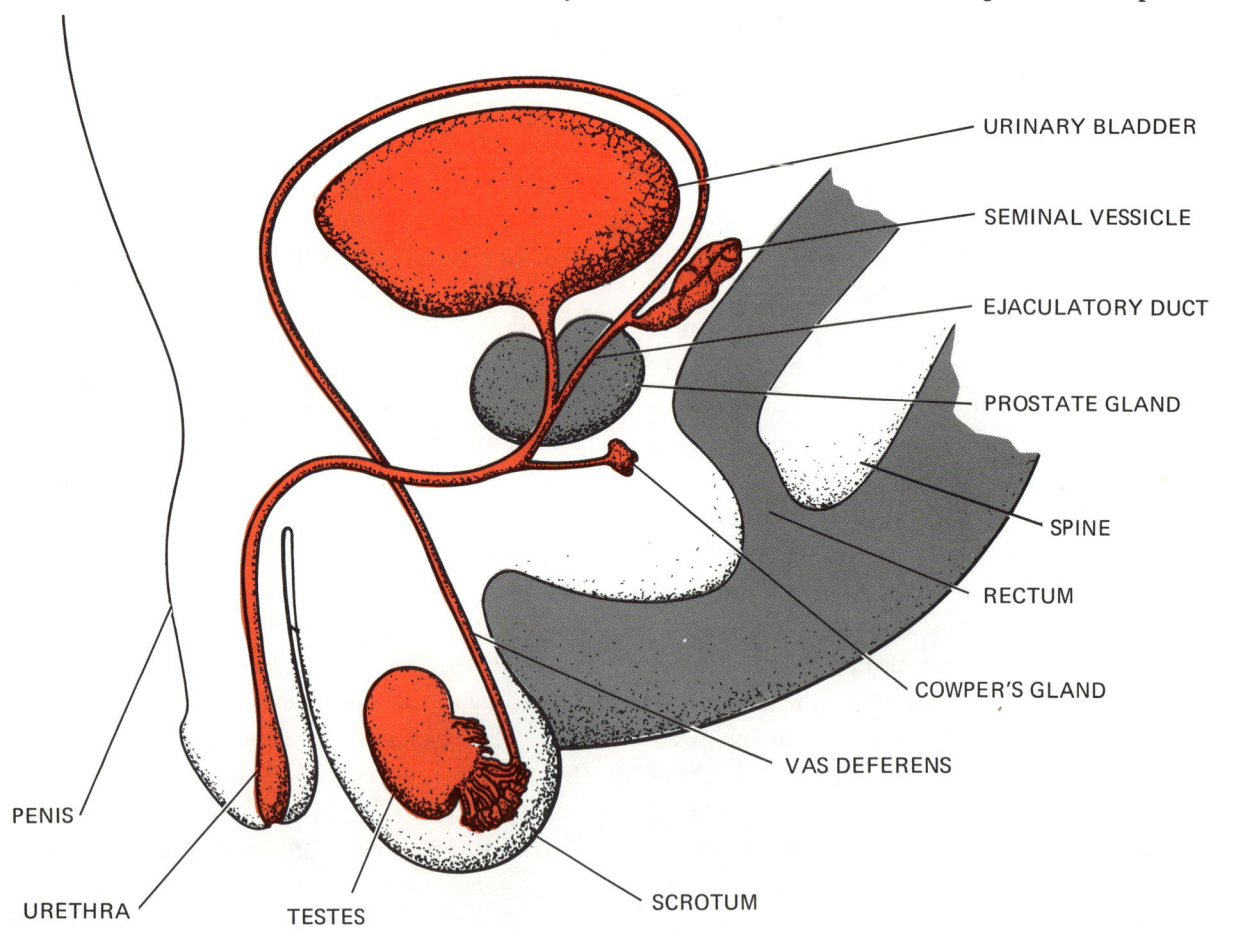

Fig. 35-3 The male reproductive organs

production starts shortly after puberty and continues until old age. Each testis produces millions of sperm. From the testes the sperm passes through a system of ducts consisting of the epididymis, the vas deferens, the ejaculatory duct, and the urethra. Starting in the epididymis, certain secretions are added to the sperm along the route of travel and make up the semen. In the male the urethra thus serves a dual purpose, it is the outlet for the reproductive tract and the urinary tract. The prostate gland, one of the accessory organs which produces a secretion that forms part of the semen, sometimes causes trouble in later life because it greatly enlarges in many older men. The enlargement presses on the urethra, sometimes making urinating impossible. Simple surgery can remedy this condition if done in time. If not, abdominal surgery may be required.

The *bulbourethral glands,* also known as Cowper's glands, are located on either side of the urethra, below the prostate gland. They add an alkaline secretion to the semen which helps the sperm to live longer than in an acid medium.

The external organs are the scrotum and the penis. Internally the scrotum is divided into two sacs, each sac containing a testis, epididymis and lower part of the vas deferens. The penis contains erectile tissue which becomes enlarged and rigid during intercourse. A loose-fitting skin covering called the *foreskin,* or *prepuce,* covers the penis. The foreskin can be removed in a simple operation known as *circumcision.*

SUGGESTED ACTIVITIES

- Discuss the relationship between ovulation, menstruation, and fertilization.
- Visit a prenatal clinic in your community. What do the examinations of the pregnant woman include on her first visit? On return visits?
- Discuss how oral and intrauterine contraceptives prevent pregnancy.
- Two blood tests which have been used to detect early pregnancy are the A-Z test and HCG test. How do they differ? What do the letters stand for?
- Investigate the value of the new microsurgery in ectopic or tubal pregnancies.

REVIEW

A. Select the letter of the item which most correctly completes the statement.

1. One of the male hormones is
   a. progesterone
   b. luteinizing hormone
   c. follicle-stimulating hormone
   d. testosterone

2. Ovulation usually occurs
   a. the day before the menstrual period begins
   b. one week before the menstrual period begins
   c. three weeks before the menstrual period begins
   d. two weeks before the menstrual period begins

HUMAN REPRODUCTION

3. The ovaries contain
   a. thirty graafian follicles
   b. thousands of graafian follicles
   c. hundreds of graafian follicles
   d. six graafian follicles

4. The development of the follicle and release of the ovum are under the influence of
   a. the follicle-stimulating hormone and the luteinizing hormone
   b. estrogen and corpus luteum
   c. progesterone and the follicle-stimulating hormone
   d. estrogen and the luteinizing hormone

5. Which one of the following statements is not correct?
   a. The fallopian tubes are about four inches long.
   b. The fallopian tubes serve as ducts for the ovum on its way to the uterus.
   c. The fallopian tubes are not attached to the ovaries.
   d. The fallopian tubes are attached to the ovaries.

B. Match each of the terms in column I with its correct statement in column II.

Column I	Column II
_____ 1. scrotum	a. secondary sex characteristics
_____ 2. testosterone	b. external sac which holds the testes
_____ 3. facial and pubic hair	c. excreted from the pituitary gland
	d. formed in the seminiferous tubules
_____ 4. epididymis and penis	e. male gamete
	f. secondary reproductive organs
_____ 5. gonadotrophic hormone	g. male hormone produced in the testes

114

unit 36 disorders of the reproductive system

OBJECTIVES

Studying this unit should help the student:

- List some common disorders of the reproductive system.
- Recognize symptoms of some common disorders.

The nurse should be familiar with the names and symptoms of the more important disorders of the reproductive system. Some of the more common ones are listed.

FEMALE REPRODUCTIVE SYSTEM

Carcinoma is a cancer which may occur in the breasts or uterus. Symptoms are not noticeable in early stages. Surgery may be necessary for either a benign or malignant tumor. *Hysterectomy* is the surgical removal of the uterus. *Mastectomy* is the surgical removal of a breast. Regular checkups help detect cancer in its early stages.

Dysmenorrhea is a term used for painful menstruation. It is characterized by abdominal pain, headache, backache, and sometimes nausea and vomiting.

Endocervicitis is inflammation of the muscle membrane lining of the cervix. The main sign is a vaginal discharge (leukorrhea).

Fibroid tumors are fibrous tumors in the uterus. They are usually benign. They may or may not show such symptoms as backache and abnormal bleeding from the uterus.

Retroversion is the backward displacement of the uterus. There may be no symptoms; or backache, constipation, or dysmenorrhea may result.

Salpingitis is inflammation of the fallopian tubes. It is accompanied by lower abdominal tenderness and pain.

Sterility is the inability to reproduce; it may occur in either sex.

MALE REPRODUCTIVE SYSTEM

Epididymitis is a painful swelling in the groin and scrotum due to infection.

Orchitis is the inflammation of a testis; it may be a complication of mumps, influenza, or other infection. A symptom is the swelling of scrotum, accompanied by elevated temperature.

Prostatitis is an inflammation of the prostate gland. By pressing on the bladder, the prostate gland causes frequent, painful urination. If pressure on the urethra is severe, urinary retention may result. A *prostatectomy* is the surgical removal of all or part of the prostate gland.

SUGGESTED ACTIVITIES

- A woman of forty-seven asks your advice concerning feelings of depression which she has had for several months. She has not been sleeping well because she perspires excessively during the night. How would you advise her?
- What is the Papanicolaou test? What is its value?

REVIEW

Match each of the terms in column I with its correct statement in column II.

Column I	Column II
_____ 1. carcinoma	a. surgical removal of uterus
_____ 2. endocervicitis	b. inflammation of fallopian tubes
_____ 3. epididymitis	c. painful swelling in groin and scrotum
_____ 4. hysterectomy	d. backward displacement of uterus
_____ 5. mastectomy	e. may cause urine to be retained
_____ 6. orchitis	f. surgical removal of breast
_____ 7. prostatitis	g. inability to reproduce
_____ 8. retroversion	h. inflammation in the lining of cervix
_____ 9. salpingitis	i. cancer in breasts or uterus
_____ 10. sterility	j. complication of disease or inflammation
	k. painful menstruation

section 8 human reproduction
SELF-EVALUATION

A. Match each of the terms in column I with its correct statement in column II.

Column I	Column II
____ 1. DNA	a. specialized sex cell of either sex
____ 2. fertilization	b. male sex cell
____ 3. gamete	c. conception
____ 4. gonadotrophic hormone	d. determine hereditary characteristics
____ 5. graafian follicle	e. reproduction division
____ 6. meiosis	f. microscopic sac
____ 7. ovum	g. female sex cell
____ 8. progesterone	h. stimulates testes to action
____ 9. sperm	i. a fertilized cell
____ 10. zygote	j. prevents menstruation during pregnancy
	k. study of genetics

B. Complete the following statements.

1. The tubes which receive the ova and allow them to pass into the uterus are the _____ .

2. The cavity below the abdominal cavity is the _____ .

3. The substance called _____ carries the inherited characteristics in the chromosomes.

4. The unborn baby formed after the zygote divides is called a(an) _____ .

5. The union of the ovum and sperm cell is called _____ .

SECTION 9

REGULATORS OF BODY FUNCTIONS

unit 37 introduction to the endocrine system

OBJECTIVES

Studying this unit should help the student:

- List the glands which make up the endocrine system.
- Describe how these glands affect body activities.
- Locate the endocrine glands in the body.

Endocrine glands are organized groups of tissues which use materials from the blood or lymph to make new compounds called *hormones*. Endocrine glands are also called ductless glands and glands of internal secretion because the hormones are secreted directly into the bloodstream as the blood circulates through the gland. The secretions are then transported to all areas of the body where they have a special influence on cells, tissues, and organs.

One of the glands (the pancreas) has two main functions. The pancreas acts as a digestive gland in the production of *pancreatic fluid* which passes through ducts to the digestive tract. Special groups of cells called islets of Langerhans secrete the hormone *insulin* which is discharged directly into the bloodstream.

① PINEAL BODY
② PARATHYROID GLANDS
③ ADRENAL GLANDS
④ ISLETS OF LANGERHANS (pancreas)
⑤ OVARIES (female gonads)
⑥ TESTES (male gonads)
⑦ THYROID GLAND
⑧ PITUITARY GLAND

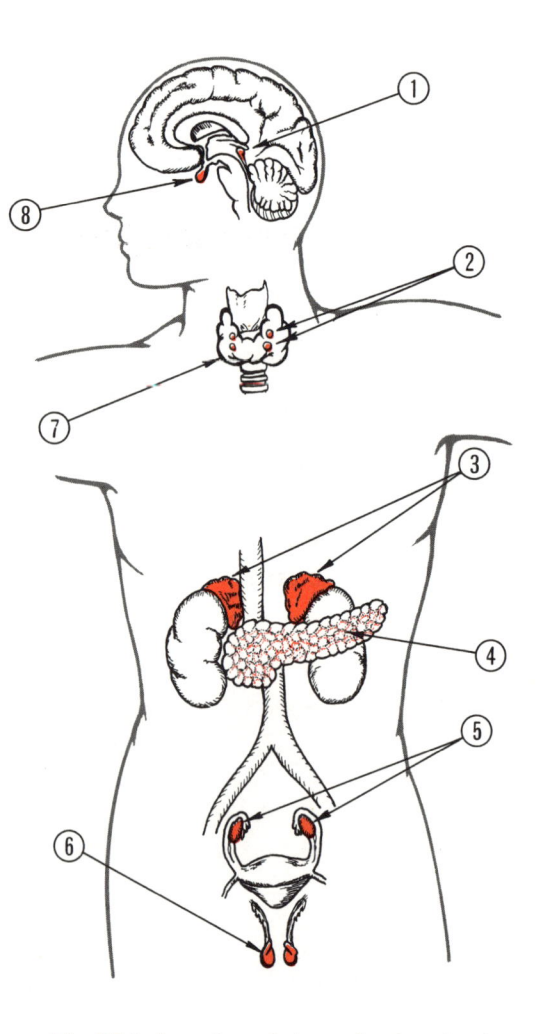

Fig. 37-1 Location of the endocrine glands

118

There are six important endocrine glands, or groups of glands, in the body.

- Pituitary gland at the base of the brain
- Thyroid gland in the neck
- Parathyroid glands near the thyroid gland
- Pancreas behind the stomach
- Two adrenal glands, one over each kidney
- Gonads, or sex glands; ovaries in the female lower abdomen, testes in the male scrotum.

Figure 37-1 shows the locations of the endocrine glands in the body. Each has specific functions to perform. Any disturbance in the functioning of these glands may cause changes in the appearance or functioning of the body. Sometimes both conditions arise.

SUGGESTED ACTIVITIES

- Discuss location, appearance, and inter-related activities of the endocrine glands.
- Do some outside reading to find out which secretions of the endocrine glands man has been able to make artificially (synthesize).
- A boy of four has shown no growth increase since he was two years of age. He seems to have a low mentality. He usually holds his mouth open; his tongue is large. His hair is dry and coarse. Which gland might be responsible for this condition?
- Discuss other hormone secretions: gastrin, secretin, placental, pineal, and thymus.

REVIEW

1. How do endocrine glands differ from other types of glands?

2. Give the general name of a secretion from an endocrine gland.

3. Name two secretions released from the pancreas.

4. Name the six important endocrine glands of the body.

5. What gland contains the islets of Langerhans?

unit 38 pituitary gland

OBJECTIVES

Studying this unit should help the student:

- Locate the pituitary gland.
- Describe the functions of the pituitary gland.
- List the principal secretions of the pituitary gland.

The *pituitary* gland is located at the base of the brain within the *sella turcica*, a small bony depression in the sphenoid bone of the skull. It is called the master gland because it secretes several hormones into the bloodstream which affect other endocrine glands. These glands, together with the other pituitary gland secretions, do much in helping maintain proper body functioning.

The pituitary gland is responsible for the growth of the long bones; thus, it controls the height of the individual. Circus giants are often the result of overgrowth of the long bones, caused by oversecretion by the pituitary gland. The organs of reproduction are influenced by it as its secretions are essential to pregnancy and lactation. It is responsible for maintaining the water balance of the body and affects the use of starches and sugars by the body. It secretes ACTH, one of the hormones used in the treatment of arthritis. The pituitary gland is also known as the *hypophyseal* gland. It is the size of a pea and consists of the *anterior* and *posterior* lobes.

It has been established that six hormones are discharged from the anterior lobe and two from the posterior. Two others appear to be discharged from an intermediate area between the two lobes. The six established hormones of the anterior lobe and the two of the posterior lobe are listed with their known functions in figure 38-2.

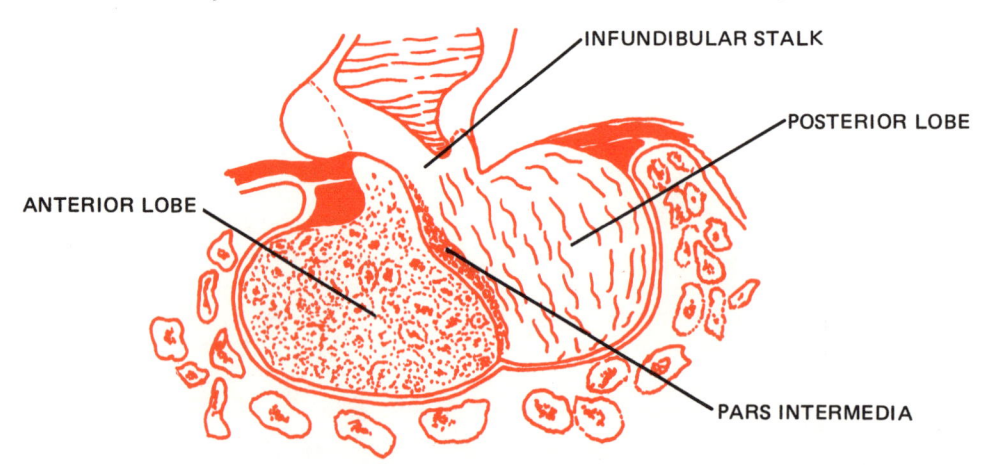

Fig. 38-1 The pituitary gland

PITUITARY GLAND

PITUITARY HORMONE	KNOWN FUNCTION
Anterior Lobe	
TSH — Thyroid-Stimulating Hormone (Thyrotropin)	Stimulates the growth and the secretion of the thyroid gland
ACTH — Adrenocorticotrophic Hormone	Stimulates the growth and the secretion of the adrenal cortex
FSH — Follicle-Stimulating Hormone	Stimulates growth of new graafian (ovarian) follicle and secretion of estrogen by follicle cells in the female and the production of sperm in the male
LH — Luteinizing Hormone (female)	Stimulates ovulation and formation of the corpus luteum
ICSH — Interstitial Cell-Stimulating Hormone (male)	Stimulates testosterone secretion
LTH — Lactogenic Hormone (Prolactin or luteotropin)	Stimulates secretion of milk and influences maternal behavior
GH — Growth Hormone (Somatotropin, STH)	Accelerates body growth
Posterior Lobe	
VASOPRESSIN (Antidiuretic Hormone, ADH)	Maintains water balance by reducing urinary output. It acts on kidney tubules to reabsorb water into the blood more quickly
OXYTOCIN	Promotes milk ejection and causes contraction of the smooth muscles of the uterus

Fig. 38-2 Pituitary hormones and their known functions

SUGGESTED ACTIVITIES

- Obtain a sheep's head from the butcher or slaughterhouse. Have the lower jaw and tongue removed. Remove the floor of the skull carefully. The pituitary gland will be found in a bony encasement called the sella turcica.
- If a fresh or preserved specimen is not available, use a plastic model of the brain or a large wall chart to locate the pituitary gland.

REVIEW

A. Briefly answer the following questions.

1. Why is the pituitary gland called the master gland?

2. Describe the principal functions of the pituitary or hypophyseal gland.

3. Why may a hormone be thought of as a chemical messenger?

REGULATORS OF BODY FUNCTIONS

B. Briefly answer the following questions.

1. What is the name of the small depression in the sphenoid bone within which the pituitary gland is located?

2. Name the two lobes of the pituitary gland.

3. How does the pituitary gland control height in an individual?

4. Give the functions of the two hormones secreted from the posterior lobe of the pituitary gland.

5. What is the approximate size of the pituitary gland?

C. Match each of the terms in column I with its correct statement in column II. Some statements are repeated.

Column I	Column II
_____ 1. thyroid-stimulating hormone	a. stimulates secretion of estrogen in the female and sperm production in the male
_____ 2. luteinizing hormone	b. stimulates testosterone secretion
_____ 3. follicle-stimulating hormone	c. stimulates growth and secretion of the adrenal cortex
_____ 4. lactogenic hormone	d. maintains water balance through kidney reabsorption
_____ 5. interstitial cell-stimulating hormone	e. stimulates secretion of milk in the mother
_____ 6. adrenocorticotrophic hormone	f. accelerates body growth
_____ 7. growth hormone	g. stimulates both the growth and secretion of the thyroid gland
_____ 8. somatotropin	h. causes contraction of smooth muscle in the uterus
_____ 9. thyrotropin	
_____ 10. prolactin	

unit 39 the thyroid and parathyroid glands

OBJECTIVES

Studying this unit should help the student:

- Locate the thyroid, parathyroid, and thymus glands.
- Describe the important functions of the thyroid gland.
- Describe the functions of the parathyroid and thymus glands.

The *thyroid* gland is located in the anterior portion of the neck, on either side of the trachea, figure 39-1. It is about two inches long and has two lobes joined by an isthmus. The thyroid has a rich blood supply. The thyroid-stimulating hormone (TSH) from the anterior lobe of the pituitary gland regulates the activity of this gland. The secretion of the thyroid gland is the hormone *thyroxin,* which stimulates the metabolic rate in the cells. This gland helps to regulate the rate of physical growth, mental development, sexual maturity, and the distribution and exchange of water and salts in the body. It can speed up or slow down the activities of the body as needed. Iodine is stored in the thyroid gland because it is essential for the manufacture of its hormone, thyroxin.

The *parathyroid* glands, usually four in number, are tiny glands the size of a grain of rice; they are attached to the posterior surface of the thyroid lobes. The only known function of these glands is control of the use of calcium and phosphorus in the body.

The *thymus* gland is located under the breastbone, or sternum. It is fairly large during childhood but begins to disappear at puberty. Little is known about the gland, but it is believed that some kind of thymus hormone is secreted from this organ during infancy; this stimulates the lymphoid cells which are responsible for the production of antibodies against some diseases.

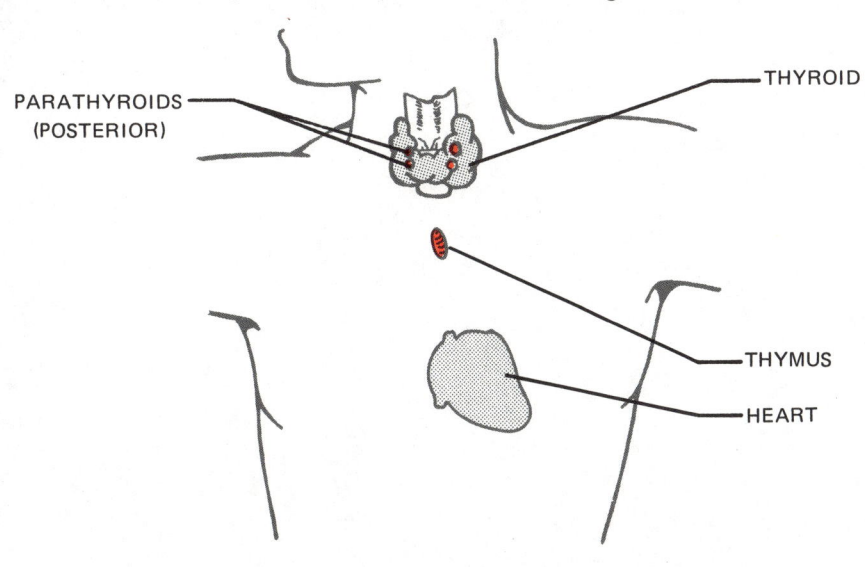

Fig. 39-1 Location of thyroid, parathyroid, and thymus glands

REGULATORS OF BODY FUNCTIONS

SUGGESTED ACTIVITIES
- Locate the thyroid, parathyroids and thymus glands using classroom models or wall charts.
- Relate the function of the thyroid gland to the process of metabolism.

REVIEW

1. Locate and describe the thyroid gland.

2. Locate and describe the parathyroid glands.

3. Describe the functions of the thyroid, parathyroid, and thymus glands.

4. Label the diagram below.

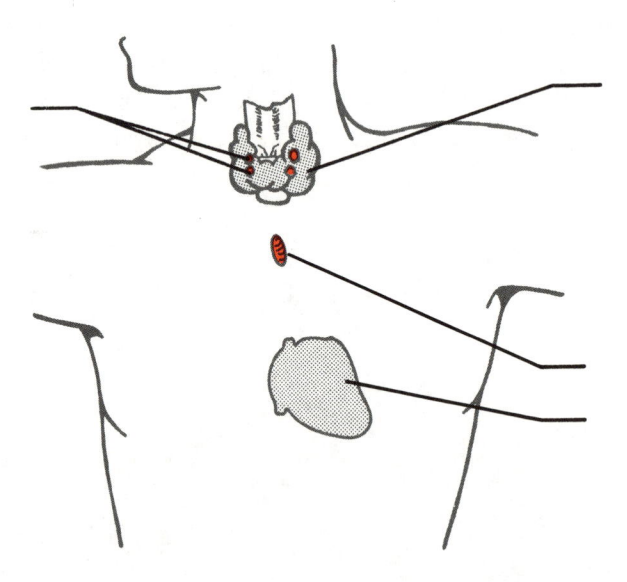

unit 40 functions of adrenal glands and gonads

OBJECTIVES

Studying this unit should help the student:

- Locate the adrenal glands and gonads.
- Describe functions of the adrenals and gonads.
- Name the secretions of the adrenals and gonads.

One of the two *adrenal* glands is located on top of each kidney, figure 40-1. Each gland has two parts, the *cortex* and the *medulla*. Adrenocorticotrophic hormone (ACTH) from the pituitary gland stimulates the activity of the cortex of the adrenal gland. The hormones secreted by the adrenal cortex are known as *corticoids*.

The cortex secretes three groups of corticoids each of which is of great importance:

1. Mineralcorticoids or (M-Cs) affect the kidney tubules by speeding up the reabsorption of sodium into the blood circulation and increasing the excretion of potassium from the blood. They also speed up the reabsorption of water by the kidneys. M-C (Aldosterone) is used in the treatment of Addison's Disease to replace deficient secretion of M-Cs.

2. The glucocorticoids (G-Cs) increase the amount of glucose in the blood. This is presumably done by (1) conversion of the protein brought to the liver into glycogen, followed by (2) breakdown of the glycogen into glucose. These G-Cs also help the body resist the aggravations caused by various everyday stresses. Both cortisone and hydrocortisone are G-Cs. The corticoids are very effective as anti-inflammatory drugs.

3. Androgens, or male sex hormones, together with similar hormones from the gonads, bring about male characteristics. They also promote protein anabolism and body growth.

The medulla of the adrenal gland secretes *epinephrine* and *norepinephrine*. Epinephrine (generic name), or Adrenalin (trade), is a powerful cardiac stimulant. It functions by bringing about a release of more glucose from glycogen for muscle activity and by increasing the force and rate of the heartbeat; this increases cardiac output and venous return and raises the systolic blood pressure.

The gonads, or sex glands, include the ovaries in the female and the testes in the male. The ovaries are located in the pelvic

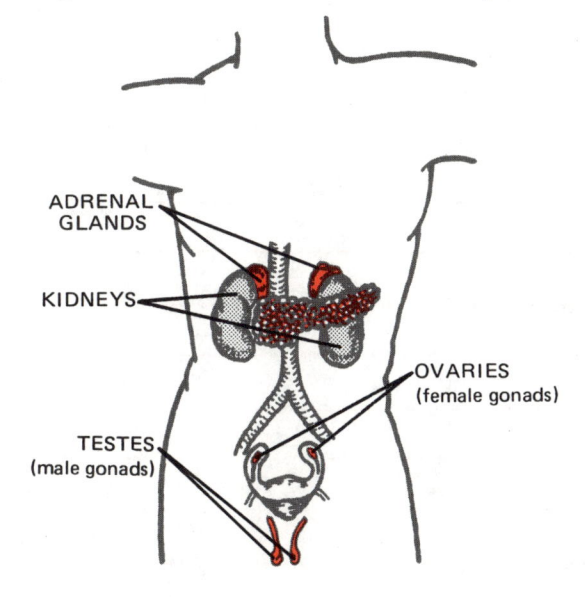

Fig. 40-1 Location of adrenals and gonads

REGULATORS OF BODY FUNCTIONS

cavity, one on either side of the uterus. The testes are located in the scrotum.

The secretions of the ovaries, *estrogen* and *progesterone,* are necessary for ovulation and the characteristic female appearance. The secretion of the testes, *testosterone,* is essential to the development of the male sex characteristics. The gonads are responsible for fertility and reproduction in both sexes.

SUGGESTED ACTIVITIES

- There are many instances on record of individuals performing superhuman feats during emergencies. Perhaps you have had such an experience yourself. What is the explanation for such strength at these times?
- Secure a lamb kidney with the adrenal gland attached. Make a long incision through the adrenal gland. Notice that it has an outer area and an inner area. Each part contributes its own secretion. Compare the lamb's adrenal gland with a picture or model of the human gland, noticing shape, size, and color.
- If frogs are available to you for laboratory work, inject about 9.2 ml of adrenalin chloride (1:100,000 solution) into the ventricle of the heart of an anesthetized frog. Count the heartbeat before the injection and after the injection. What did the adrenalin do to the heart? What was the quality of the heartbeat before and after injection?
- Describe the use of adrenalin for hemorrhage and asthma conditions.

REVIEW

1. Describe the location of the adrenal glands.

2. Summarize the functions and secretions of the adrenals and the gonads, using the following chart.

GLAND		SECRETION	FUNCTION
Adrenals	Cortex		
	Medulla		
Gonads	Ovaries		
	Testes		

unit 41 the pancreas

OBJECTIVES

Studying this unit should help the student:

- Locate the pancreas.
- Explain endocrine functions of the pancreas.
- Explain the body's need for insulin.

The *pancreas* is an organ located behind the stomach. The gland cells of the pancreas are concerned with the production of pancreatic juice, a digestive juice. The islet cells secrete the hormone *insulin.* Therefore, the pancreas is a gland of external secretion and internal secretion.

The islet cells (mostly beta cells) are distributed throughout the pancreas. These cells were named the islets of Langerhans after the doctor who discovered them. Beta cells produce insulin which: (1) promotes the utilization of glucose in the cells, necessary for maintenance of normal levels of blood glucose; (2) promotes fatty acid transport into cells and fat deposition in them; (3) promotes amino acid transport into cells; and (4) promotes protein synthesis. Lack of insulin secretion by the island (islet) cells causes diabetes mellitus.

SUGGESTED ACTIVITIES

- If laboratory facilities are available, obtain a sheep or beef pancreas. Prepare a thin section for use with the microscope.
    a. Focus the slide under low power to show the arrangement of the islets. Discuss what you see.
    b. Focus the slide under high power to see the arrangement of the cells in the islets and the pancreas in general.

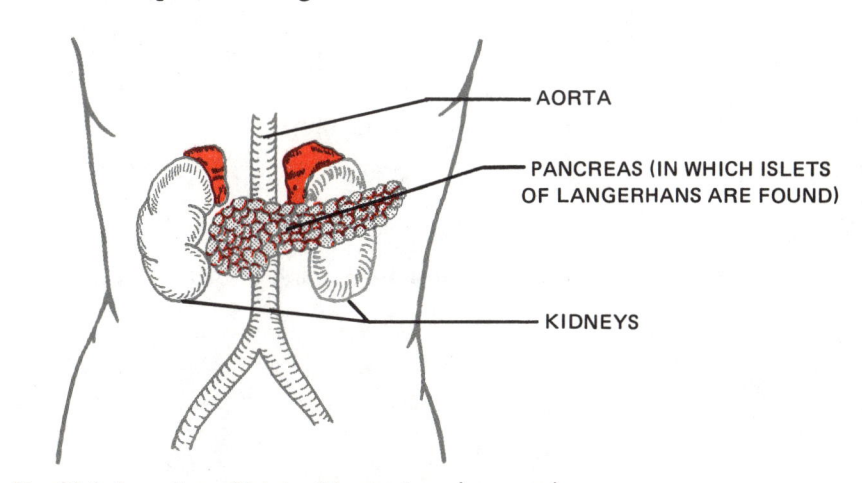

Fig. 41-1 Location of islets of Langerhans (pancreas)

- Obtain several specimens of urine to test for the presence of sugar. Pour 5 milliliters (cc) of Benedict's solution into a test tube. Using a medicine dropper, add 10 drops of urine to the solution. Shake well to mix thoroughly. Boil the solution in a test tube in water bath for 3-5 minutes. Note the color.

  a.
COLOR	INDICATION
Blue	Sugar free
Green	+ present, trace
Greenish yellow	1+
Yellow	2+
Brown or red	3+ to 4+

  b. Your instructor may have new tablet testing devices for the presence of sugar. Try these simplified methods as well as the Benedict's solution method.

- Research and discuss whether insulin may be considered a cure for diabetes.

REVIEW

1. What is meant by the statement that the pancreas is a gland of external as well as internal secretion?

2. Where is the pancreas located?

3. Where in the pancreas is the hormone insulin made?

4. What are the functions of insulin?

5. What condition is caused by a lack of insulin in the bloodstream?

unit 42 endocrine gland disorders

OBJECTIVES

Studying this unit should help the student:
- List causative factors of endocrine gland disorders.
- Recognize some endocrine gland disorders which interfere with body functions.
- Relate treatment to some of the more common types of gland disorders.

Endocrine gland disturbances may be caused by several factors such as disease of the gland itself, infections in other parts of the body, and dietary deficiencies. Most disturbances result from (1) hyperactivity of the glands, causing oversecretion of hormones, or (2) hypoactivity of the gland resulting in undersecretion of hormones.

THYROID DISTURBANCES

Hyperthyroidism is due to overactivity of the thyroid gland; too much thyroxin is produced and the gland becomes enlarged. There is an increase in the basal metabolic rate, usually accompanied by marked prominence of the glands, bulging of the eyeballs, rapid heartbeat, hand tremors, and irritability.

Determination of the basal metabolism rate (BMR) is an aid in diagnosing thyroid functioning. The test is performed by measuring the amount of oxygen the person uses at rest; from this, his basal metabolism rate can be calculated. Another diagnostic test is the protein-bound iodine test (P.B.I.). This is a blood test which determines the concentration of thyroxin in the bloodstream. Other thyroxin level tests are the T_3 and T_4. These tests are more accurate than the P.B.I. but also are more difficult to perform.

The radioactive iodine uptake test measures the activity of the thyroid gland. Dilute radioactive iodine is given by mouth. The amount accumulated in the thyroid gland is calculated by use of a Geiger counter. *Simple goiter* is an enlargement of the thyroid gland due to a deficiency of iodine in the diet.

Hypothyroidism is a condition in which the underactivity of the thyroid gland slows down all body processes. Hypothyroidism may lead to either cretinism or myxedema. *Cretinism* is a condition which develops in infancy or early childhood in which mental and physical development is retarded. *Myxedema* is a condition similar to cretinism, but developing in adulthood.

The treatment for thyroid deficiency usually consists of giving the patient enough thyroid extract to bring metabolism up to normal. If a child is given the thyroid product early in life, he will usually develop normally, overcoming the tendency to low mentality and growth deficiencies such as dry skin, thick tongue, or coarse hair.

PARATHYROID DISTURBANCES

The parathyroid glands regulate the use of calcium and phosphorus. Both of these minerals are involved in many of the body systems.

Hyperfunctioning of the parathyroid glands may cause an increase in the amount of blood calcium, thereby increasing the tendency for the calcium to crystallize in the kidneys as *kidney stones*. Excess amounts of calcium and phosphorus are withdrawn from the bones; this may lead to eventual deformity.

Hypofunctioning of the glands may cause *tetany*. This is a disease marked by convulsive twitchings. Vitamin D and calcium are given to restore normal balance.

PITUITARY DISTURBANCES

Disturbances of the pituitary gland may produce many body changes. It is chiefly involved in the growth function; but as the master gland, the pituitary indirectly influences other activities.

Hyperfunctioning of the gland may cause overgrowth of the long bones leading to *gigantism* or, in adulthood, to *acromegaly*, an overdevelopment of the bones of the hands, feet, and face.

Hypofunctioning of the gland may cause *dwarfism, diabetes insipidus,* and *menstrual disturbances*.

ADRENAL DISTURBANCES

Overactivity of the adrenal gland may result in virilism, Cushing's syndrome or aldosteronism. Underactivity may give rise to disorders like Addison's disease; this condition is manifested by a drop in blood sugar, blood pressure, blood volume because of the serious imbalance of electrolytes.

GONAD DISTURBANCES

Disturbances in the ovaries may consist of *cysts* and *tumors, abnormal menstruation,* and *menopausal changes*. Turner's syndrome may occur in either the male or female; this is a chromosomal disorder.

PANCREAS DISTURBANCES

Diabetes mellitus is a condition caused by decreased secretion of insulin from the islet cells of the pancreas in which carbohydrate metabolism is disturbed. This has an adverse effect on protein and fat metabolism.

SUGGESTED ACTIVITIES

- Discuss how a disturbance in the pituitary gland can affect the functioning of other endocrine glands.
- Discuss some of the symptoms of diabetes mellitus.

REVIEW

1. What is the meaning of each of the following terms.
   a. hyperfunctioning

   b. hypofunctioning

2. What symptoms indicate hyperfunctioning of the thyroid gland?

3. What conditions may result from hypofunctioning of the pituitary gland?

ENDOCRINE GLAND DISORDERS

4. Complete the chart which follows.

GLAND	NORMAL FUNCTION	DISTURBANCES
Pituitary		
Thyroid		
Parathyroids		
Thymus		
Adrenals		
Gonads		
Pancreas		

section 9 regulators of body functions
SELF-EVALUATION

A. Match each of the terms in column I with its correct statement in column II.

Column I	Column II
_____ 1. ACTH	a. master gland of the endocrine system
_____ 2. adrenals	b. any gland of internal secretion
_____ 3. cortisone	c. a hormone secreted by adrenals
_____ 4. gonad	d. regulates use of calcium
_____ 5. endocrine	e. the secretion of any endocrine gland
_____ 6. hormone	f. helps body meet emergencies
_____ 7. insulin	g. sex gland
_____ 8. parathyroid	h. regulates body metabolism
_____ 9. pituitary	i. one of the hormones secreted by pituitary gland
_____ 10. thyroid	j. a hormone which regulates carbohydrates and metabolism
	k. hypofunction of endocrine glands

B. Label the glands numbered below.

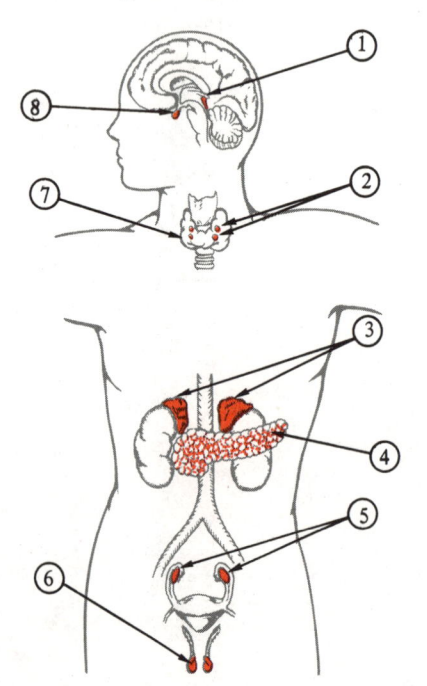

1. _____
2. _____
3. _____
4. _____
5. _____
6. _____
7. _____
8. _____

SECTION 10

COORDINATION OF BODY FUNCTIONS

unit 43 introduction to the nervous system

OBJECTIVES

Studying this unit should help the student:

- Describe the functions of the nervous system.
- List the main parts of the nervous system.
- Describe three types of neurons.
- Define characteristics of the nerve cells.

The study of body functions reveals that the body is made up of millions of small structures that carry on a multitude of different activities; these are coordinated and integrated into one harmonious whole. The two main communication systems are the endocrine system and the nervous system. They send chemical messengers and nerve impulses to all of the structures. The endocrine system and hormonal regulation have been discussed previously. Hormonal regulation is slow, while neural regulation is comparatively rapid.

The nervous system is the most highly organized system of the body and consists of the brain, spinal cord, and nerves. The structural and functional unit, as in other systems, is the cell. The nerve cell, or *neuron,* is especially constructed to carry out its function of communication. In addition to the nucleus, cytoplasm, and cell membrane, the neuron has extensions of cytoplasm from the cell body. These extensions, or processes, are called *dendrites* and *axons.* There may be several dendrites, but only one axon. These processes, or fibers, as they are often called, are paths along which impulses travel, figure 43-1.

All neurons possess the characteristics of being able to react when stimulated and of

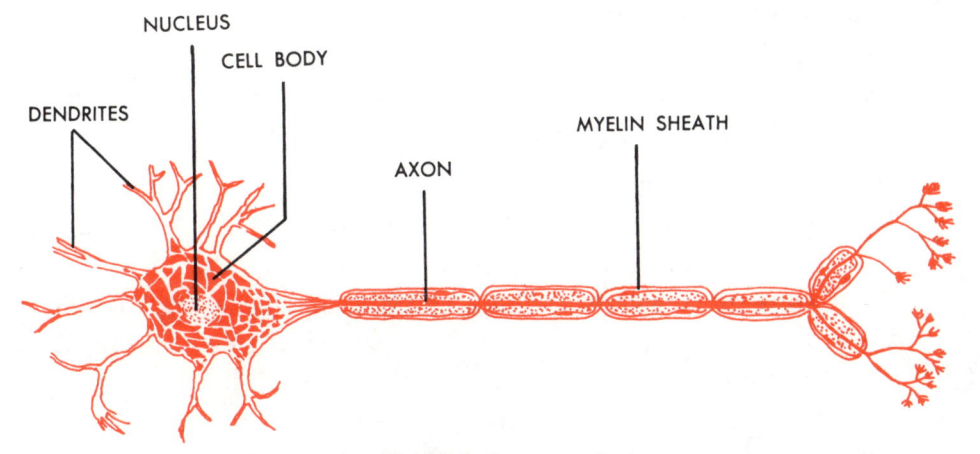

Fig. 43-1 A nerve cell

133

COORDINATION OF BODY FUNCTIONS

being able to pass along the stimulus to other neurons. These characteristics are *irritability* (the ability to react when stimulated) and *conductivity* (the ability to transmit a disturbance to distant points). The dendrites receive the impulse and transmit it to the cell body, and then to the axon where it is passed on to another neuron or to a muscle or gland.

There are three types of neurons:

1. Sensory neurons which emerge from the skin or sense organs and carry messages, or impulses, toward the spinal cord and brain
2. Motor neurons which carry messages from the brain and spinal cord to the muscles and glands
3. Connecting, or association, neurons which carry impulses from one neuron to another

The nervous system can be divided into three divisions:

1. The central nervous system which consists of the brain and spinal cord
2. The peripheral nervous system which is made up of the nerves of the body consisting of twelve pairs of cranial nerves extending out from the brain and thirty-one pairs of spinal nerves extending out from the spinal cord
3. The autonomic nervous system which consists of a chain of ganglia (group of neuron cell bodies) on either side of the spinal cord

Where decision is called for and action must be considered, the central and peripheral nervous systems are involved. They carry information to the brain where it is interpreted, organized, and stored, and a command is sent to organs or muscles. The autonomic nervous system supplies heart muscle, smooth muscle, and secretory glands with nervous impulses as needed. It is usually involuntary in action.

SUGGESTED ACTIVITIES

- Look up and learn the meaning of each of the following terms:

| receptors | stimulus | reflex | ganglia |
| effectors | response | impulse | synapse |

REVIEW

Complete the following statements.

1. Two types of muscle tissue supplied with nerve impulses from the autonomic nervous system are _____ and _____ .
2. Neurons which emerge from the skin or sense organs and carry messages toward the spinal cord are called _____ neurons.
3. Neurons which carry messages from the brain and spinal cord to the muscles and glands are called _____ neurons.
4. The ability of a neuron to react when stimulated is called _____ .
5. The extension of the nerve cell body which receives the impulse is called the _____ and the part which passes the impulse on is called the _____ .
6. _____ glands receive stimulation from the autonomic nervous system.

unit 44 the central nervous system: brain and spinal cord

OBJECTIVES

Studying this unit should help the student:

- List the parts of the brain.
- Describe the structure of the brain and spinal cord.
- Relate functions to parts of the brain.

The brain is a soft mass of nervous tissue. It weighs about three pounds and is protected by the bony skull. The brain is covered by three membranes called *meninges*. The innermost covering, directly over the brain tissue is the *pia mater*. It carries blood vessels to the brain and spinal cord. The middle layer is the *arachnoid*. The space between the arachnoid and pia mater is filled with cerebrospinal fluid, which is made in the ventricles and acts as a liquid shock absorber. The *dura mater* is the outer meninge; it is a tough protective membrane which lines the brain on one side and the underside of the skull on the other.

Brain tissue is made up of gray and white matter. The outer surface of the brain is grayish and the center, white. The *cortex* is the highest center of the brain and is made up of

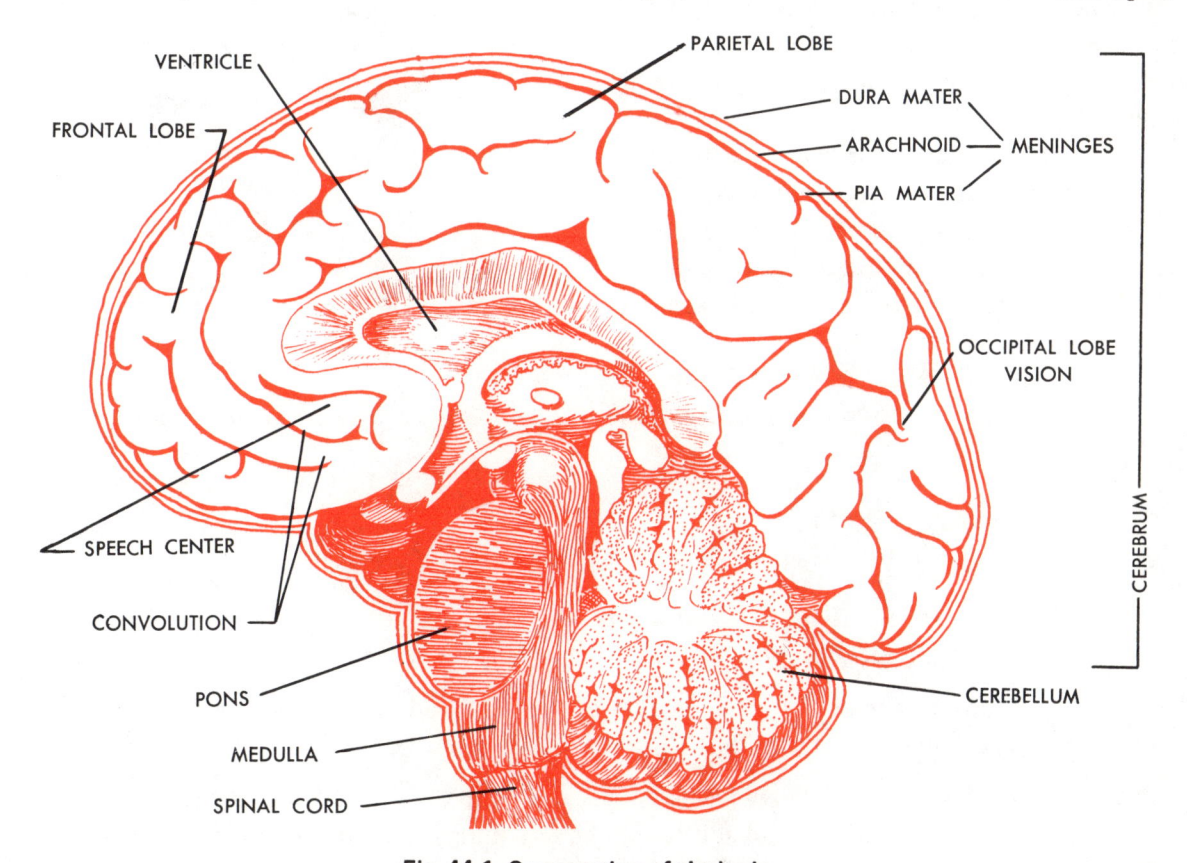

Fig. 44-1 Cross section of the brain

135

COORDINATION OF BODY FUNCTIONS

so-called gray matter. The gray matter really consists of millions of nerve cell bodies and naked nerve fibers; the white matter contains millions of nerve cell fibers with myelin sheaths, which accounts for the difference in appearance.

The brain is divided into three parts: the *cerebrum, cerebellum,* and *brain stem.* The brain stem is further divided into several parts, two of which are the *medulla,* and *pons.*

The cerebrum is the largest part of the brain. Its surface is covered with wrinkles, causing the brain to have the appearance of little mounds (convolutions) separated by grooves. The *convolutions* serve to increase the surface area of the brain.

The cerebrum is divided into right and left hemispheres; it contains the centers of reasoning, memory, thought, speaking, and sensation. Cells in the right hemisphere control voluntary movements which occur in the left side of the body. The left hemisphere controls voluntary movements of the right side of the body.

The cerebellum is much smaller than the cerebrum, and it, too, is divided into two hemispheres. The function of the cerebellum is to coordinate the muscular movements of the body and thus govern the steadiness of these movements.

Beneath the cerebral hemispheres are the *thalamus* which relays messages and the *hypothalamus* which helps regulate body temperature, water balance, and appetite. The *medulla* includes the centers which regulate heartbeat, respiration, swallowing, hiccoughing, and coughing.

The spinal cord continues down from the medulla. It is white and soft and lies within the vertebrae of the spinal column. Like the brain, it is submerged in cerebrospinal fluid and is surrounded by the three meninges. In the spinal cord, the gray matter is located in the internal section and the white matter composes the outer part, figure 44-2. In the gray matter of the cord, connections can be made between incoming and outgoing nerve fibers which provide the basis for reflex action. The spinal cord functions as a reflex center and as a conduction pathway to and from the brain.

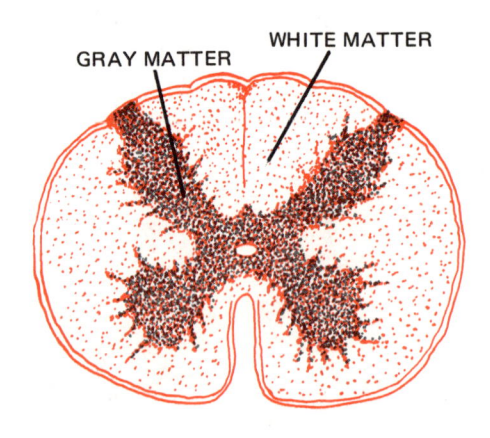

Fig. 44-2 Cross section of the spinal cord

SUGGESTED ACTIVITIES

- If laboratory facilities are available, make arrangements to observe the dissection of the brain of a sheep or calf. Locate the cerebrum, cerebellum, medulla, convolutions, hemispheres.

- From memory, draw a diagram of the brain and identify the structures. Compare it with figure 44-1 after you have finished it.

- Make up a chart which lists the parts of the brain and spinal cord. Do outside research and add functions which may not have been covered in this unit.

THE CENTRAL NERVOUS SYSTEM: BRAIN AND SPINAL CORD

REVIEW

Match each of the terms in column I with its correct statement in column II.

Column I	Column II
_____ 1. meninges	a. the innermost covering of brain tissue
_____ 2. cortex	b. separates the other two meninges
_____ 3. arachnoid	c. matter which contains naked nerve fibers
_____ 4. cerebellum	d. fluid made in the ventricles of the brain
_____ 5. pia mater	e. largest of 3 parts of the brain
_____ 6. white matter	f. nerve fibers covered with myelin sheath
_____ 7. cerebrum	g. part of the brain which coordinates movements
_____ 8. gray matter	h. the highest center of the brain
_____ 9. dura mater	i. the outer meninge
_____ 10. cerebrospinal fluid	j. regulates heartbeat, respiration, and coughing
	k. three membranes that cover the brain

unit 45 the peripheral and autonomic nervous systems

OBJECTIVES

Studying this unit should help the student:

- Explain how a simple reflex act is carried out by the nervous system.
- Define the terms associated with reflex action.
- Relate the functions of the sympathetic and parasympathetic nervous systems.

The peripheral nervous system includes all the nerves of the body. The autonomic nervous system is a specialized part of the peripheral system and controls the involuntary or automatic activities of the vital internal organs.

A nerve is composed of bundles of nerve fibers and small blood vessels all enclosed by connective tissue. If the nerve is composed of fibers that carry impulses from the sense organs to the brain or spinal cord, it is called a *sensory,* or *afferent,* nerve; if it is composed of fibers carrying impulses from the brain or spinal cord to muscles or glands, it is called a *motor,* or *efferent,* nerve; and if it contains both sensory and motor fibers, it is called a *mixed nerve.*

Some of the twelve pairs of cranial nerves (such as the facial nerve) are mixed nerves; some (as the optic nerve) contain only sensory fibers; and some are entirely motor nerves. All thirty-one pairs of spinal nerves are mixed nerves. Each of them divides into branches which either go directly to a particular body segment or form networks with adjacent spinal nerves called *plexuses.*

The autonomic nervous system includes nerves, ganglia, and plexuses which carry impulses to all smooth muscle, secretory glands, and heart muscle. It thus regulates the activities of the *visceral organs* (heart and blood vessels, respiratory organs, alimentary canal, kidneys and urinary bladder, and reproductive organs). The activities of these organs are usually automatic and not subject to conscious control, without special training.

The autonomic system consists of two divisions: the *sympathetic* and the *parasympathetic.* These two divisions are antagonistic in their action. For example, the sympathetic system may accelerate the heartbeat in response to fear, but the parasympathetic will slow it down. Normally the two divisions are in balance; the activity of one or the other becomes dominant as dictated by the needs of the organism.

The sympathetic nervous system consists primarily of two cords which begin at the base of the brain and proceed down both sides of the spinal column. These consist of nerve fibers and ganglia of nerve cell bodies. The cord between the ganglia is a cable of nerve fibers, closely associated with the spinal cord. Sympathetic nerves go to all the vital internal organs including the sweat glands, liver, and pancreas, and to such muscles as the heart, stomach, intestines, blood vessels, the iris of the eye, and the bladder.

The parasympathetic system is composed of two important active nerves: the *vagus* nerve which extends from the medulla and

THE PERIPHERAL AND AUTONOMIC NERVOUS SYSTEMS

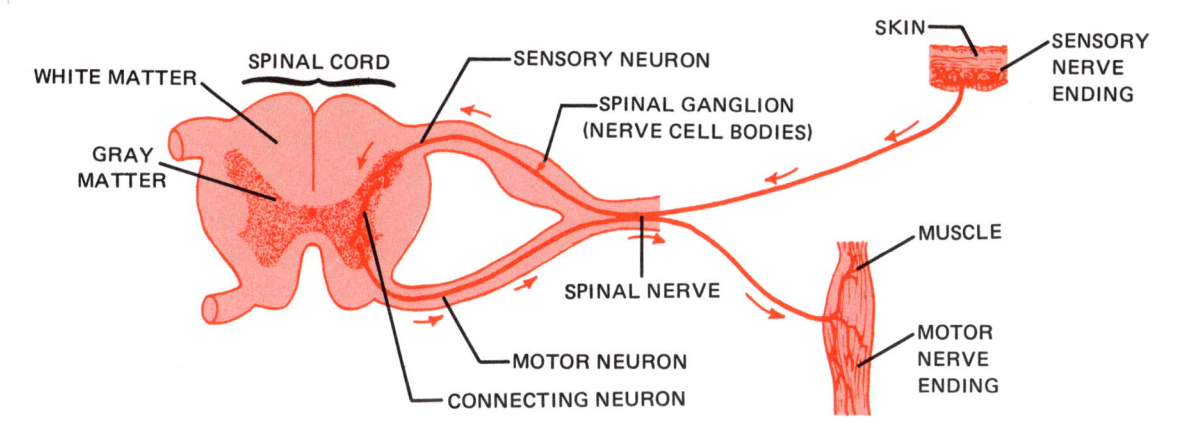

Fig. 45-1 The reflex arc

proceeds down the neck, sending branches to the chest and neck and the *pelvic* nerve which emerges from the spinal cord around the hip region and sends branches to the organs in the lower part of the body.

Both the sympathetic and the parasympathetic are greatly influenced by emotion. During periods of fear, anger, or stress, the sympathetic division acts to prepare the body for emergency action. The effects of the parasympathetic are generally to counteract the effects of the sympathetic. For example, the sympathetic system constricts the blood vessels, thus raising blood pressure, and the parasympathetic system dilates the blood vessels, thus lowering blood pressure. The two systems work as a pair, striking a nearly perfect balance when the body is functioning properly.

The Reflex Act

The simplest type of nervous response is the *reflex* act, which is unconscious and involuntary. The blinking of the eye when a particle of dust touches it, the removing of the finger from a hot object, the secretion of saliva at the sight or smell of food, the movements of the heart, stomach, and intestines, are all examples of reflex actions.

Every reflex act is preceded by a stimulus. Anything in the environment which causes activity is called a *stimulus.* Examples of stimuli are sound waves, light waves, heat energy, and odors. Special structures called *receptors* pick up these stimuli. For example, the retina of the eye is the receptor for light; special cells in the inner ear are receptors for sound waves; and special structures in the skin are the receptors for heat.

Reaction to the stimulus is called the *response.* The response may be in the form of movement, in which case the muscles are the *effectors,* or responding organs; if the response is in the form of a secretion, the glands are the effectors. Reflex actions involving the skeletal muscles are controlled by the spinal cord and involve only sensory, connecting, and motor neurons. This is called the *reflex arc,* figure 45-1.

SUGGESTED ACTIVITIES

- Discuss the action of the autonomic nervous syteem when a patient is served an attractive, appetizing lunch.
- List ten reflex acts.

COORDINATION OF BODY FUNCTIONS

REVIEW

A. Complete the following statements.

1. A nerve is composed of bundles of fibers, called _____ and small blood vessels.

2. A nerve is composed of fibers which carry impulses from sense organs to the brain or spinal cord is called a _____ or _____ nerve.

3. A nerve composed of fibers which carry impulses from the brain or spinal cord to muscles or glands is called a _____ or _____ nerve.

4. A mixed nerve contains both _____ and _____ fibers.

5. The autonomic nervous system is a specialized part of the peripheral system and controls _____ .

6. The autonomic nervous system has two parts which counterbalance each other; these are the _____ and _____ systems.

B. Identify the structures on the diagram. Enter your answers below.

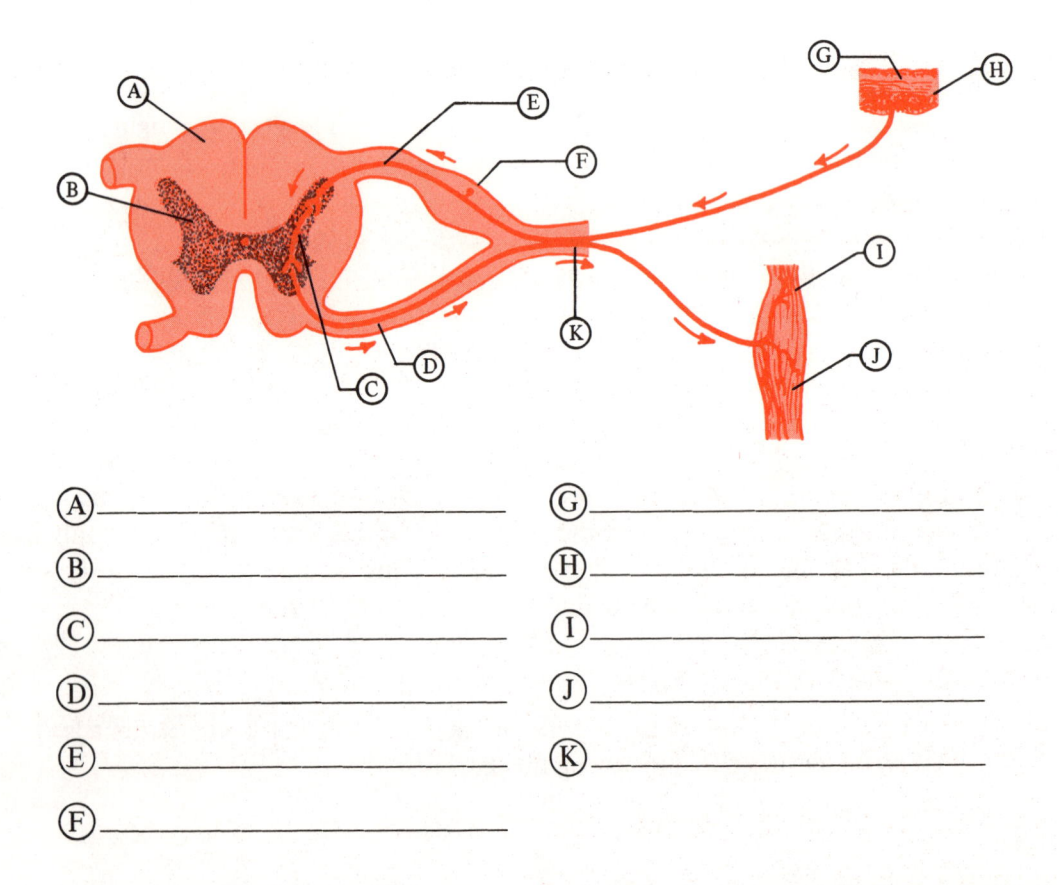

Ⓐ _____   Ⓖ _____
Ⓑ _____   Ⓗ _____
Ⓒ _____   Ⓘ _____
Ⓓ _____   Ⓙ _____
Ⓔ _____   Ⓚ _____
Ⓕ _____

140

unit 46 special sense organs-eye and ear

OBJECTIVES

Studying this unit should help the student:

- Explain how stimulation of a sense organ results in sensation.
- List the parts of the eye and relate them to their functions.
- List the parts of the ear and relate them to their functions.

Sensory receptors are special structures which are stimulated by changes in the environment. Sensory receptors (touch, pain, temperature, and pressure) are found all over the body, located either in the skin or connective tissues. Special sensory receptors include the taste buds of the tongue, special cells in the nose, the retina of the eye, and the special cells in the inner ear which make up the organ of Corti. When a sense organ is stimulated, the impulse travels along nerve pathways to the brain, where it is registered in a certain area. Sensation actually takes place in the brain but the sensation is referred back to the sense organ mentally. This is called *projection* of the sensation.

THE EYE

The eye is the sense organ which is stimulated by light rays. The wall of the eye is made up of three layers or coats: the sclera, the choroid, and the retina, figure 46-1.

The outer layer is called the *sclera* or *sclerotic* coat. It is tough in order to protect the delicate structure within. The visible portion of this coat is the so-called white of the eye. Also in the front, in the very center of the sclerotic coat, is a circular, clear area which is called the *cornea*. This is sometimes referred to as the "window of the eye." It is transparent to permit light rays to pass through it.

The middle layer of the eye is the *choroid* coat; it is pigmented and has blood vessels to

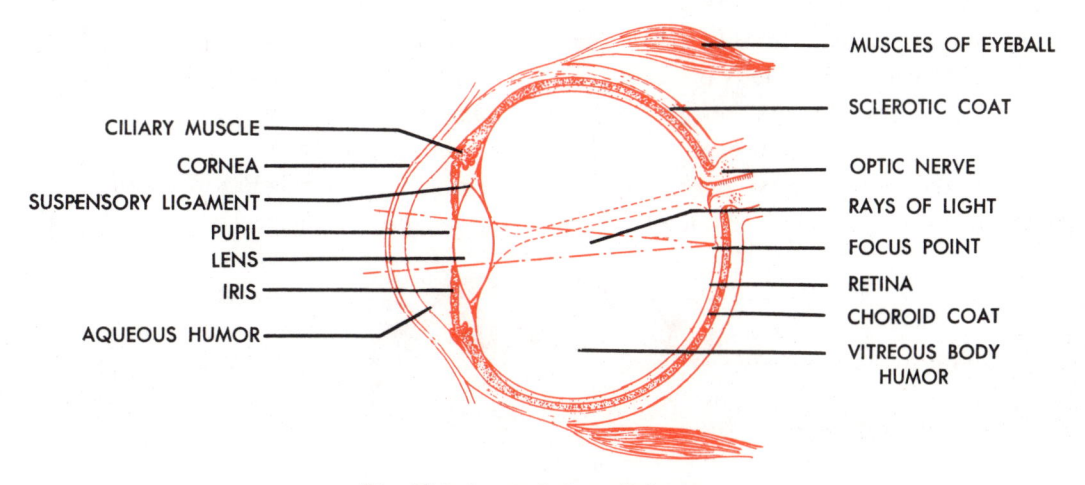

Fig. 46-1 Internal view of the eye

COORDINATION OF BODY FUNCTIONS

nourish the eye. In front, the choroid coat has a circular opening called the *pupil*. A colored, muscular ring surrounds the pupil. It is the iris, or colored part of the eye. By contraction of its muscles the iris regulates the size of the pupil and thus determines the amount of light which may enter the eye. The pupil gets smaller in bright sunshine so that the eye does not get too much light; it gets bigger in a darkened room or theater, in order to permit as much light as possible to enter. In this way, the eye may be compared to a camera; the iris corresponds to the shutter or diaphragm.

The *retina* of the eye is the innermost or third coat of the eye. It does not extend around the front portion of the eye. It is upon this sensitive layer that the light rays from an image are focused, corresponding to the film or plate in a camera. The retina contains pigment and specialized cells known as *rods* and *cones* which are sensitive to light. The part of the retina where the nerve fibers enter the optic nerve to go to the brain does not have these specialized cells, and therefore, is not sensitive to light. For this reason, it is often called the "blind spot." The *lens* is a crystal structure just behind the iris. Light rays travel through it and are bent or refracted in order that they may focus on the retina.

The *aqueous humor* is a watery fluid which fills the compartment in front of the lens. The jellylike substance which fills the compartment in back of the lens is called the *vitreous body*. Both aid in refraction of light. After the image is focused on the retina it travels via the optic nerve to the visual portion of the brain, located towards the back of the head just above the neck. If rays of light do not focus correctly on the retina, the condition can be corrected by wearing properly fitted lenses which bend the rays of light accurately.

The eyeball is moved by muscles. The eye is protected by the bone surrounding it and by the eyebrows, eyelids, and eyelashes, figure 46-2.

THE EAR

The ear is a special sense organ which is especially adapted to pick up sound waves and send these impulses to the auditory center of the brain, located in the temporal area just above the ears. The receptor for hearing is the delicate *organ of Corti* which is located within the cochlea of the inner ear.

The ear has three parts, the outer or external ear, the middle ear, and the inner ear, figure 46-3. The outer ear consists of the visible portion and a canal which leads to the ear drum (tympanic membrane).

The middle ear is really a cavity in the temporal bone. It connects with the pharynx by means of a tube called the *eustachian tube*. This tube serves to equalize the air pressure in the middle ear with that of the outside atmosphere. A chain of three tiny bones is found in the middle ear: the *hammer*, the *anvil*, and the *stirrup;* they transmit sound waves from the ear drum to the inner ear.

The inner ear consists of several membrane-lined channels which lie deep within the temporal bone. The special organ of hearing is a spiral-shaped passage known as the *cochlea* which contains a membranous

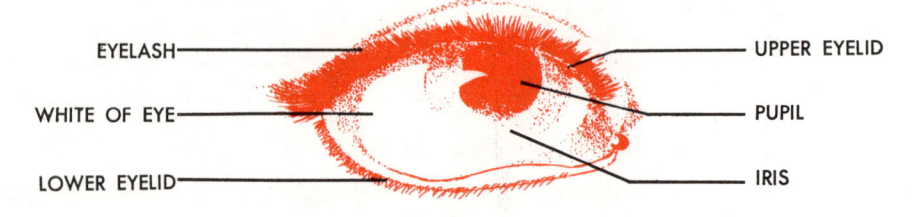

Fig. 46-2 External view of the eye

SPECIAL SENSE ORGANS — EYE AND EAR

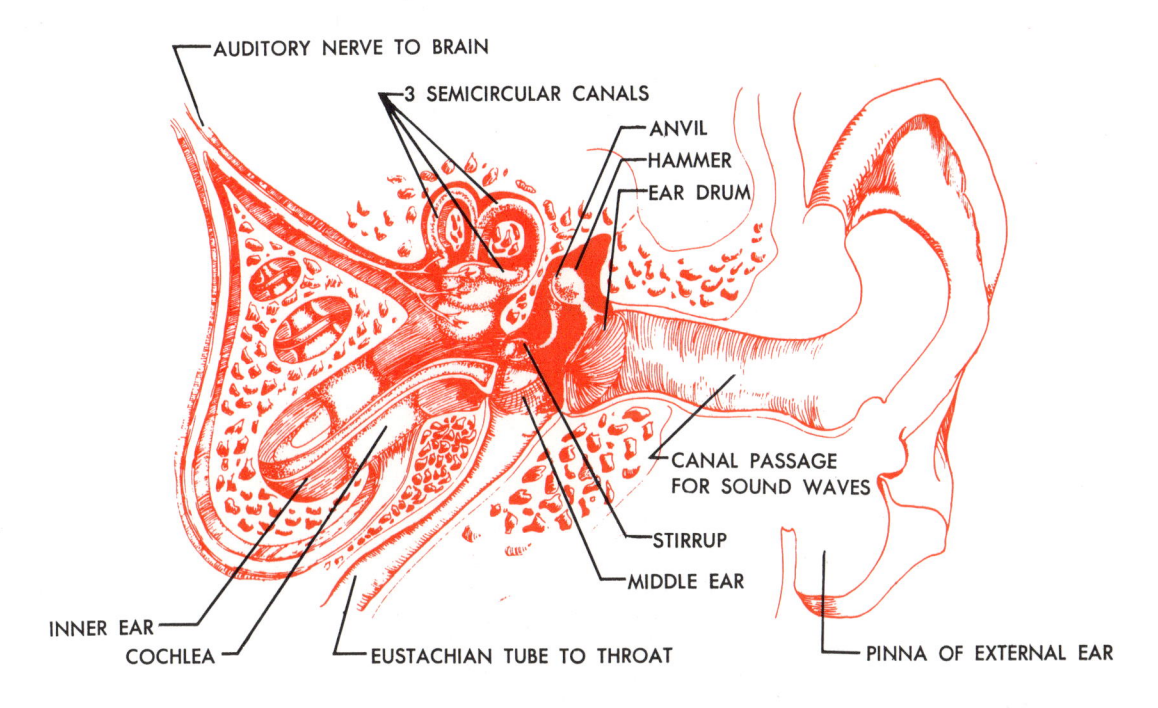

Fig. 46-3 Diagram of ear

tube called the *cochlear duct*. This duct is filled with fluid which vibrates when the sound waves from the stirrup bone hit it. Located in the cochlear duct are delicate cells which make up the organ of Corti. These hairlike cells pick up the vibrations caused by the sound waves against the fluid and transmit them through the auditory nerve to the hearing center in the brain.

Three *semicircular canals* also lie within the inner ear. They contain a liquid and delicate hairlike cells which bend when the liquid in the canals is set in motion by head and body movements. These impulses are sent to the brain and help maintain body balance, or equilibrium. They have nothing to do with the sense of hearing.

SUGGESTED ACTIVITIES

- Using an eye model, identify the chief structures in the eye. Explain their functions.
- Using an ear model, identify the chief structures in the ear. Explain their functions.
- Using a diagram or a chart of the skin, identify the four receptors found there.
- Locate the taste buds on the tongue, using a chart.
- In figure 46-3, trace the path of sound waves from the time they strike the eardrum until the sensation of hearing is registered in the brain.
- Using figure 46-1, trace the path of light rays from the time they strike the cornea until the sensation of sight is registered in the brain.

COORDINATION OF BODY FUNCTIONS

REVIEW

1. Name the three structural coats found in the eye, giving the function of each one.

2. Explain how refraction of light is important to vision.

3. Describe how the organ of Corti functions.

4. Identify the structures in the following diagram.

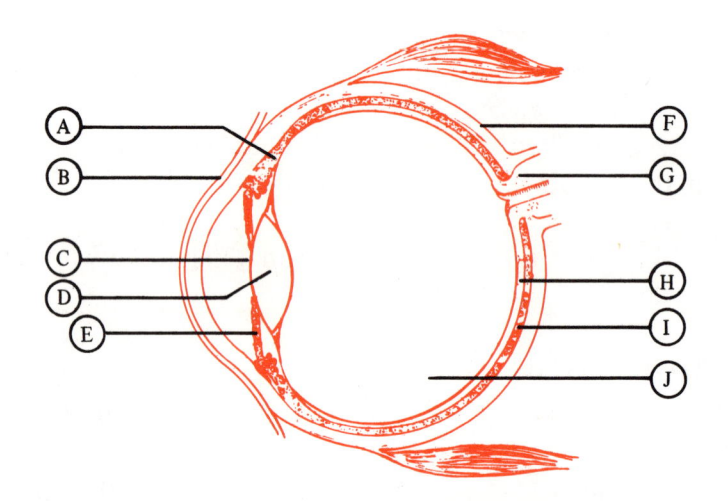

Ⓐ _____   Ⓕ _____
Ⓑ _____   Ⓖ _____
Ⓒ _____   Ⓗ _____
Ⓓ _____   Ⓘ _____
Ⓔ _____   Ⓙ _____

144

unit 47 diseases of the nervous system

OBJECTIVES

Studying this unit should help the student:

- Recognize symptoms of some common ailments of the nervous system.
- Relate treatment to some common nervous system diseases.
- List common ear and eye disorders and related treatment.

The nurse must become familiar with symptoms and treatments for disorders of the nervous system. Some of the more common ones are mentioned here.

NERVOUS SYSTEM DISORDERS

Chorea, or *St. Vitus' Dance,* which is characterized by involuntary twitching of the muscles of the legs, arms, and face, usually occurs in children. The disease may last from three to six months. Treatment consists of rest, nourishing food, and protection from excitement of any kind.

Shingles or *herpes zoster,* results in eruptions on the skin, accompanied by pain along the nerves in some parts of the body. This disorder is believed to be caused by a virus infection of nerve endings. The involved area, often chest or abdomen, must be treated by protecting it from air and from the irritation of clothing.

Neuralgia is a sharp and stabbing pain along the course of a nerve. It is usually a symptom of some other disease. There is no structural change in the nerve.

Neuritis is inflammation of a nerve trunk. It causes pain and may result in hypersensitivity, loss of sensation, weakness of muscles, muscle paralysis and atrophy.

Sciatica is a form of neuralgia or neuritis of the sciatic nerve in the leg.

Poliomyelitis, or *infantile paralysis,* is a viral infectious disease of the nerve pathways in the spinal cord. The muscles which are controlled by these diseased nerve paths become paralyzed. Death may occur. Treatment consists of hot packs, exercises given by a trained person, and special exercises given under water. The patient may have to be put in an "iron lung" if the muscles of respiration are involved. The Sister Kenny method of treatment is given by trained specialists. Vaccines are now available to protect against the disease. All children should be immunized against polio.

Encephalitis is a brain disease. There are several forms of encephalitis. "Sleeping sickness" is a form of the disease accompanied by drowsiness, stupor, and great weakness.

Cerebral palsy is a disturbance in voluntary muscular action due to brain damage. Definite causes are unknown. It may be due to birth injuries, intracranial hemorrhage, or infections such as encephalitis.

Hydrocephalus is an increased volume of cerebrospinal fluid within the cavity of the brain. This causes enlargement of the skull and prominence of the forehead.

Convulsions, characterized by violent muscle contractions, may occur because of high fever, lack of vitamin D, or brain tumors; they cause brain tissue to discharge abnormal nerve signals.

COORDINATION OF BODY FUNCTIONS

Acute bacterial meningitis is the infection of the meninges of the brain and spinal cord. Early diagnosis and treatment with antibiotics reduces the severity of the disease.

Epilepsy is a disease of the nervous system which may be characterized by mild or violent convulsions. The nurse cannot control or shorten the convulsion, but should protect the patient so he does not injure himself during violent movements. Medication is now available which, if taken regularly, will usually control attacks.

EAR DISORDERS

Otitis media is an infection of the middle ear. It usually causes an earache. This disorder is often a complication of the common cold in children.

EYE DISORDERS

Conjunctivitis is an inflammation of the conjunctive membranes in front of the eye. Redness and discharge of mucus occurs. Since it may be contagious, it should be promptly treated by a physician.

Glaucoma is a condition of the eye in which the aqueous humor does not circulate properly within the eye so pressure increases within the eyeball. If it is untreated glaucoma leads to blindness because it damages the retina and optic nerve. With prompt treatment, total blindness may be avoided. Early detection is most important, and treatment will usually prevent progress of the disease.

A *cataract* is a condition characterized by a lack of transparency of the lens. Light cannot pass through the clouded lens and therefore the person cannot see. The vision is corrected with eyeglasses or the opaque lens is removed by surgery and a synthetic lens replacement made.

A *sty* is a tiny abscess at the base of an eyelash. It is due to inflammation of one of the sebaceous glands of the eyelid.

SUGGESTED ACTIVITIES

- Discuss the procedure the nurse should follow in caring for a patient seized by an epileptic attack. What observations should be reported?
- Prepare for a panel discussion on rehabilitation of cerebral palsy patients. Include family and social problems.
- Discuss how deafness can be helped or treated.
- Discuss the reason for the infrequency of attacks of poliomyelitis today.

REVIEW

A. Match each of the terms in column I with its correct statement in column II.

Column I	Column II
_____ 1. acute bacterial meningitis	a. disease which may be characterized by convulsions
_____ 2. poliomyelitis	b. infection of the brain membranes
_____ 3. epilepsy	c. middle ear infection
_____ 4. glaucoma	d. viral disease for which there is now a vaccine
_____ 5. cataract	e. a clouded or opaque lens
	f. condition caused by increased pressure within the eyeball

B. Answer the following questions.
1. What is the difference between neuralgia and neuritis?

2. What is sciatica?

3. What disease results when the membranes of the brain become inflamed?

4. What are the hazards of untreated glaucoma?

section 10 coordination of body functions
SELF-EVALUATION

A. Identify each numbered part and give its function.

① _____

② _____

③ _____

④ _____

⑤ _____

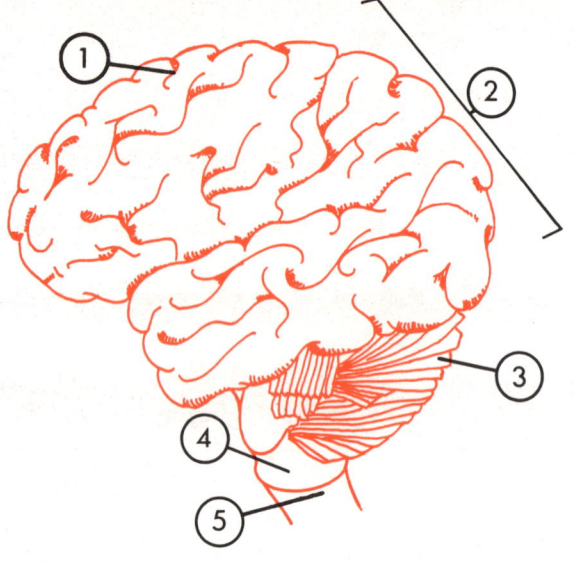

B. Using the diagram of the reflex arc, trace the path of the impulse. Identify the numbered parts and describe the action.

① _____

② _____

③ _____

④ _____

⑤ _____

⑥ _____

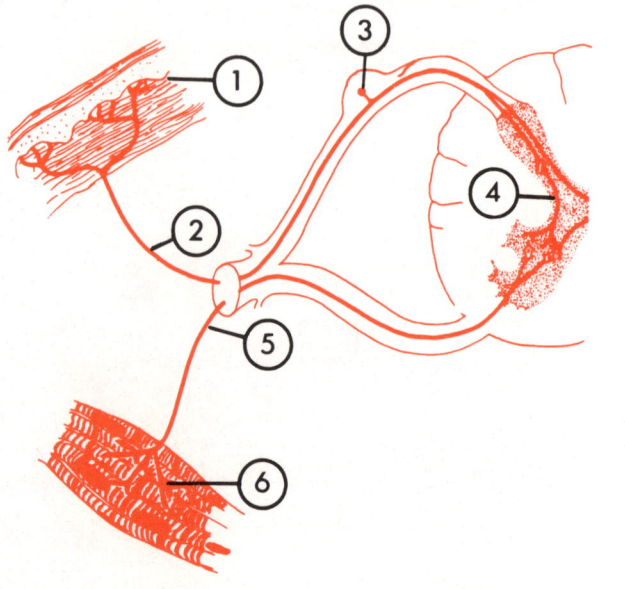

SELF-EVALUATION

C. Place the correct answers in the blank spaces.

1. The unit of structure in nervous tissue is the _____.

2. There are two chief types of neurons, the _____ or _____ which carries messages to the brain and the _____ or _____ which conveys messages away from the central nervous system to peripheral areas.

3. Axons of nerve cells have a protective covering around them called the _____.

4. The brain and spinal cord are covered by three _____. They are, from the center out, _____, _____, and _____.

5. The brain includes the _____, _____, and _____.

6. The spinal cord lies within the _____.

7. Connections between the brain and surrounding structures are established by _____ nerves.

8. Connections between the spinal cord and other structures are established by _____ nerves.

9. The _____ is the largest part of man's brain.

10. The four receptors for sensation which are found in the skin are those for touch, cold, heat, and _____.

SECTION 11

CARE OF THE BODY FUNCTIONS

unit 48 practices for health maintenance

OBJECTIVES

Studying this unit should help the student:

- Identify health practices which affect each of the body systems.
- Explain why these specific health practices contribute to good health.
- Describe how to care for the feet, eyes, ears, and hair.
- State why exercise and rest are important for good health.

Studying body structure and functions usually creates concern about the everyday practices needed to maintain the body. Developing good health habits works to the benefit of all body systems.

GETTING ENOUGH EXERCISE AND REST

Exercise makes a person breathe more deeply and helps the heart, arteries, lungs, and other internal muscles maintain a normal state. When the muscles are exercised, lymph flows through the body; this aids in the nutrition of the tissues and the removal of wastes. By stimulating the muscles of the digestive tract, exercise helps to push the food along the intestinal tract. In this way, constipation may be relieved by exercise. Exercise should not, however, reach the point of extreme fatigue.

Relaxation and amusement afford relief from the physical and nervous tension which may accompany work. Mental fatigue can be worse than physical fatigue. The brain should not be forced to work when it is already tired. Hobbies and forms of recreation which do not tax the brain and nerves should be indulged in by those whose work involves continuous mental activity. Gardening, swimming, boating, walking and many other such activities offer relaxation and health-giving exercise.

Sleep is of utmost importance. It is an accepted fact that sleep is necessary for maintaining both physical and emotional health, although the exact metabolic or physiologic need has not been fully explained. (Chronic loss of sleep may result in personality changes.) Children need more sleep than adults because growth is taking place. Eight or nine hours of sleep are advised for the average adult person in order to keep his system in the best possible functional state. The mental health of an individual is of tremendous importance. The mental, emotional, moral, and spiritual health of an individual all contribute to his physical well-being.

CARE OF THE FEET

Because the feet bear the weight of the body, well-fitting shoes should be worn. The foot has a close arrangement of twenty-six bones. Many of these bones are delicate and slender and if they are crowded into small or ill-fitting shoes, injury and discomfort are inevitable. Corns, bunions, and fallen arches are some of the conditions people suffer as a result.

Properly fitted shoes should have the heel low and broad with the inner edge of the shoe almost straight. The toe of the shoe must provide sufficient room to permit all the toes to be separate, not cramped together. The selection of shoes should be guided by the advice of a podiatrist, if necessary. Health workers and others who must be on their feet a good portion of the day should wear professionally fitted shoes.

Flat feet and fallen arches should be referred to a podiatrist for treatment. Toe and foot exercises are beneficial to arches and stimulate circulation.

THE HEART AND CIRCULATION

Heart disease takes more lives annually in the United States than any other illness. Medical authorities are of the opinion that anxiety and the competition of modern living are causative factors in heart attacks.

Obesity places an added burden on the heart and blood vessels, increasing susceptibility to heart and circulatory ailments. Loss of weight plus rest and peace of mind is essential to the treatment of people who have circulatory diseases. Eliminating animal fats and substituting polyunsaturated oils in the diet is an unproved step toward lowering the blood cholesterol which seems to be a factor in coronary heart disease. Relaxation before meals is indicated. Going from a taxing job to the dinner table and eating large meals rapidly may contribute to heart attacks.

New drug therapy in the form of antihypertensive drugs and tranquilizing medications has proved to be a factor in allaying anxieties and in contributing to the decrease in the incidence of coronary occlusions. The concerned individual is seeking closer medical supervision, following the dietary regime specified by the physician, keeping within his normal weight range, resting adequately, getting the prescribed amount of daily exercise, and learning a more positive philosophy of life.

FRESH AIR AND BREATHING

Proper ventilation and correct breathing are necessary for good health. Breathing through the nostrils warms the air and filters out the dust. The habit of mouth breathing may be the result of adenoid enlargement or other nasal obstructions. Correct breathing involves expanding the lungs as much as possible; deep breathing means there is more air to fill the lungs. Correct posture is essential to proper breathing.

Good ventilation provides for a constant source of fresh air. Windows should be opened from the bottom to permit fresh cool air to enter, and from the top to allow stale, warm air to leave a room. Air in heated rooms in winter is often dry and should be moistened by having a receptacle of water in some part of the room.

Within the past few years there has been a marked increase in the incidence of respiratory ailments in such conditions as emphysema and cancer of the lung. Although repeated and chronic bronchitis may be a basic causative factor in emphysema, there is little doubt that smoking and air pollution are predisposing causes. Although cancer of the lung and cancer in general may have some genetic origin, public health authorities are exerting great effort to overcome air pollution in urban centers and to discourage cigarette smoking.

DEVELOPING GOOD EATING HABITS

A diet should consist of nourishing food, chosen from the basic four food groups,

CARE OF THE BODY SYSTEMS

well-prepared, and attractively served. Diet should be modified, however, to meet the changes that age makes in nutritional needs.

Since lifetime eating habits are often set in youth, proper diet at this time is doubly important. An adequate supply of proteins and calories seems to have direct correlation with size and mental ability. Therefore, the nourishment the fetus receives, the diet provided for the young child and the example of good eating habits set for him by the adults he watches, all contribute to his physical and mental development toward adulthood.

Adolescence is typified by rapid growth and a large appetite. It is important that nutritional requirements are met during this time of great physical activity. Later, during middle age, when there is less physical activity, caloric intake must be cut to prevent an undesired weight gain. Old age, also, is a time of decreased activity and may require a change in diet and eating habits. Because nutrition at this time must meet individual health needs, diet should be discussed with and directed by a doctor.

Food should be chewed thoroughly and slowly to enable the digestive juices to penetrate the particles of food for proper digestion. Eating between meals is usually not recommended for a normal, healthy person. Under special circumstances, when a person may have some particular ailment requiring more frequent feedings, strict adherence to the regime is necessary.

Anger, worry, fear, and rigorous physical or mental activity before or after meals tends to interfere with proper digestion. Eating should be a pleasant experience.

Foods which contain the minerals, calcium and phosphorus are essential for strong teeth. Milk, eggs, green vegetables, and whole grains contain these minerals. Fish liver oils and egg yolks (which contain vitamin D) are also necessary for good teeth. Citrus fruits contain vitamin C and benefit teeth and gums. Some hard food should be chewed daily to exercise the gums. Brushing the teeth after every meal and making semiannual visits to the dentist help maintain teeth in the best condition possible. Fluoridation of the water supply is recommended by health authorities to prevent tooth decay. Diseased teeth are not only a source of discomfort but they may lead to more serious health problems.

PRACTICING PERSONAL HYGIENE

The daily bath is important in removing dirt and keeping the skin fresh and clean. Taking a warm bath, using a mild, unscented soap, relaxes a tired body and helps induce sleep. A cool shower tones the muscles and stimulates the circulatory and nervous systems.

The hair should be brushed every morning and evening in order to stimulate the circulation of the blood in the scalp. It should be kept clean by frequent shampooing. Hands should be washed before every meal, before preparing food, and after using the bathroom. Cleanliness, fresh clean garments, deodorants, neatness, and good grooming are all important assets to one's well-being.

It is as important to wash the feet daily as it is to clean any other part of the body. Wastes are excreted through the perspiration in the feet and, since air does not circulate readily around the feet to evaporate the moisture, bacteria tends to produce bad odors unless daily bathing care is practiced.

The diet should include food fiber in order to provide bulk to help the digestive process. Proper elimination is necessary for the healthy functioning of the body. Regular bowel movements are aided by proper diet, exercise, freedom from tension, and a sufficient intake of liquids: six to eight glasses of water should be drunk daily. Foods which contain bulk such as cereals, fruits, and vegetables help to prevent constipation. Laxatives should be unnecessary.

Plenty of water also helps to flush out the kidneys and replace liquid lost in perspiration. Constant backaches in the small of the back and burning or irritation during urination should be promptly reported to a physician. A routine urinalysis is an important part of every physical examination.

CARE OF THE EYES AND EARS

The human eye is a delicate structure; yet it is subjected to all kinds of abuses. Reading in poor light or at twilight when the light is fading may produce eyestrain. Reading when very bright light or sunlight is directly facing the eyes may be harmful. If light is reflected from a very shiny surface, reading should be avoided. The best light for reading is provided when the light falls over the left shoulder on to the page. Long, continued use of the eyes in sewing, reading fine print, or performing any other close work, can produce eyestrain unless the eyes are rested occasionally. This can be done by looking off into space or to the side. Reading in moving vehicles or when lying down may also cause a strain on the eyes.

At the first sign of discomfort such as headache, squinting, a sty, and blurred vision, a doctor should be visited. No drops of any kind should be instilled without having the physician check the eyes to determine the cause of the discomfort. An *ophthalmologist* is a medical doctor who specializes in eye disorders. He will prescribe eye drops, glasses, or contact lenses if they are necessary. He will also determine if the eye problem is due to some other condition rather than an eye disorder.

The ear can be injured easily. An applicator or any other object introduced too far into the ear canal can cause serious injury. Also, the dry cotton may adhere to the wall and cause an inflammatory process. The ears may be gently cleaned with the moistened end of a washcloth. However, any foreign object which becomes lodged in the ear should be removed only by a physician.

SUGGESTED ACTIVITIES

- Plan an educational public health campaign which aims at reducing the incidence of cases of emphysema and cancer of the lung in the area in which you live.
- Prepare a library report on the average life span of man today compared to that of man fifty years ago. Explain why this change has come about.
- A man who is overweight has been ordered by his physician to follow a low-cholesterol diet. Make a list of those foods which he should eliminate from his diet and another list of foods which he can eat.

REVIEW

1. Relate choice of diet and eating habits to the control of heart disease.

2. Give two reasons for the importance of a proper diet in early childhood.

CARE OF THE BODY SYSTEMS

3. Why do children require more sleep than adults?

4. Name two practices which contribute to good foot care.

5. What does brushing the hair do in relation to maintaining health?

6. Drinking plenty of water performs what two functions?

7. List four conditions under which reading produces eyestrain.

8. Explain how exercise aids digestion.

9. How can windows be positioned to provide proper ventilation of a room?

10. What advantage does nose breathing have over mouth breathing?

GLOSSARY

abdomen: portion of the body lying between the thorax and pelvis

abduct: to draw away from the midline

absorption: passing of a substance into body fluids and tissues

acetabulum: cup-shaped cavity in the innominate bone receiving the head of the femur

Achilles tendon: cord at the back of the heel

acidosis: disturbance in the acid-base balance from excess acid or excessive loss of bicarbonate; depletion of alkaline reserve

adduct: move toward the midline. The opposite of abduct.

albuminuria: excess of albumin in the urine

alkalosis: excessive alkali; disturbance in the acid-base balance from excess loss of acid

amenorrhea: absense of the menses

aneurysm: a widening or sac formed by the dilation of a blood vessel

angina: disease marked by painful attacks of tightness, choking or suffocation

ankylosis: abnormal immobility and consolidation of a joint

anorexia: loss of appetite

antibody: a substance produced by the body that inactivates a specific foreign substance which has entered the body

antigen: a substance which stimulates the formation of antibodies against it

anus: outlet from the rectum

aphasia: loss of the ability to speak; also may be accompanied by loss of verbal comprehension

arachnoid: weblike middle membrane of the meninges

areola: pigmented ring around the nipple

arteriole: small branch of an artery

artery: blood vessel which carries blood away from the heart

ascites: accumulation of fluid in the peritoneal cavity

ataxia: muscle incoordination

atrium: upper chamber of the heart

atrophy: wasting away of tissue

autonomic: independent or self-regulating

axilla: armpit

axon: a nerve-cell structure which carries impulses away from the cell body

biceps: large flexor muscle of the upper arm or leg

brachial: pertaining to the arm

bronchiectasis: condition in which the bronchi are dilated

bronchiole: small branch of a bronchus

buccal: pertaining to the cheek or mouth

bursa: sac or pouch lined with synovial membrane and containing fluid

calculus: stone-like formation in any part of the body; usually composed of mineral salts

calyx: cup-shaped part of the renal pelvis

capillary: microscopic blood vessel which connects arterioles with venules

carcinoma: a malignant tumor

caries: decay of a tooth or bone

carotid: refers to arteries which supply blood to the neck and head

carpal: refers to bones of the wrist

catalyst: chemical substance which alters a chemical process but does not enter into the process

cataract: condition in which the eye lens becomes opaque

cecum: pouch at the proximal end of the large intestine

cerumen: ear wax

cervix: neck; usually the rounded, conical protrusion of the uterus into the vagina

GLOSSARY

cholecystectomy: removal of the gallbladder

cholesterol: a sterol normally synthesized in the liver and also ingested in egg yolks, animal fats and tissues

chromosome: nuclear material which determines hereditary characteristics

chyme: food which has undergone gastric digestion

cilia: tiny lashlike processes of protoplasm

cochlea: spiral cavity of the internal ear containing the organ of Corti

congenital: present at birth

coronary: referring to the blood vessels of the heart

corpus: body

cortex: outer part of an internal organ

costal: pertaining to the ribs

cretinism: congenital and chronic condition due to the lack of thyroid hormone

cutaneous: pertaining to the skin

cyanosis: bluish color of the skin due to insufficient oxygen in the blood

cytoplasm: protoplasm of the cell body, excluding the nucleus

deciduous teeth: temporary teeth usually lost by six years of age

defecation: elimination of waste material from the rectum

deltoid: triangular-shaped muscle which covers the shoulder prominence and is used for intramuscular injections in adults

dendrite: nerve cell process that carries nervous impulses toward the cell body

dentin: main part of the tooth located under the enamel

dentition: number, shape, and arrangement of teeth

dermis: true skin; lying immediately beneath the epidermis

dextrose: glucose, a monosaccharide which may accumulate in the urine

diaphragm: muscular partition between the thorax and the abdomen

diastole: dilation state of the heart; the rest between systoles

digestion: change of foods into compounds that can be assimilated

dorsal or posterior: pertaining to the back

dropsy: accumulation of serous fluid in a body cavity; edema

dyspnea: labored breathing or difficult breathing

ectopic: in an abnormal position; said of an extrauterine pregnancy or cardiac beats

edema: excessive fluid in tissues

electrolytes: electrically charged particles which help determine fluid and acid-base balance

embolism: obstruction of a blood vessel by a circulating blood clot, fat globule, air bubble or piece of tissue

embryo: the human young the first three months after conception

emesis: vomitus

emphysema: lung disorder in which inspired air becomes trapped and is difficult to expire

empyema: pus in a cavity

endocrine: pertaining to a gland which secretes into the blood or tissue fluid, instead of into a duct

enzyme: organic catalyst that initiates and accelerates a chemical reaction

epidermis: outmost layer of skin

epinephrine: adrenalin; secretion of the adrenal medulla, which prepares the body for energetic action

erythrocyte: red blood cell

exophthalmos: abnormal protrusion of the eyes

fascia: band or sheet of fibrous membranes covering or binding and supporting muscles

fetus: the human young from the third month of the intrauterine period until birth

fibrin: an insoluble protein necessary for the clotting of blood

GLOSSARY

fibrinogen: a protein which is converted into fibrin by the action of thrombin
fontanel: unossified areas in the infant skull; soft spot

ganglion: a mass of nerve cell bodies outside the central nervous system
gastric: pertaining to the stomach
gene: part of the chromosome that transmits a given hereditary trait
genitals: reproductive organs, also called genitalia
gestation: development period of the human young from conception to birth
glomerulus: compact cluster of capillaries in the nephron of the kidney
glucose: a monosaccharide or simple sugar; the principal blood sugar
gluteal: pertaining to the area near the buttocks
glycerin or glycerol: product of fat digestion
glycogen: polysaccharide formed and largely stored in the liver; can be converted into glucose when needed
gonads: sex glands (ovaries or testes)

hemiplegia: paralysis of one side of the body
hemoglobin: oxygen-carrying pigment of the blood
heparin: substance obtained from the liver, which slows blood clotting
hernia: protrusion of a loop of an organ through abnormal opening
histology: microscopic study of living tissues
hormone: chemical secretion, usually from an endocrine gland
hyperopia: farsightedness
hypertension: abnormally high blood pressure
hypertrophy: enlargement of a part due to increase in size of its already existing cells

incus: the middle ear bone, also called the anvil
inguinal: pertaining to the groin

inhalation: taking air into the lungs
insulin: hormone secreted by the pancreas; regulates the rate of carbohydrate usage
involuntary: opposite of voluntary, not within the control of will
involution: return of an organ to its normal size after enlargement; also the regressive changes due to aging
ion: an electrically charged atom
irritability: ability to react to a stimulus; excitability
isotonic: the same tension or pressure

keratectomy: excision of part of the cornea
kilogram: 1000 grams or approximately 2.2 pounds
kinesthesia: ability to perceive the direction or weight of muscular movement
kinetic: pertaining to motion

labia: lips
lacrimal: pertaining to tears
lactation: secretion of milk from the breasts
lactose: milk sugar; a disaccharide used in infant formulas
lateral: toward the side
leukocyte: white blood cell
ligament: a band of fibrous tissue connecting bones or supporting organs
lipid: fatty compound
lumbar: pertaining to the loins; region between the posterior thorax and sacrum
lumen: passageway or opening to a tubular structure such as a blood vessel
lymph: watery fluid in the lymphatic vessels
lymphocyte: a type of white blood cell

malleus: largest of three middle ear bones; also called the hammer
maltose: disaccharide formed by the hydrolysis of starch
mammary: pertaining to the breast
mastication: the process of chewing
meatus: passageway or opening

GLOSSARY

mediastinum: intrapleural space; separating the sternum in front and the vertebral column behind

medulla: inner portion of an organ

membrane: a thin layer of tissue which covers a surface or divides an organ

menstruation: monthly discharge of blood from the uterus; also called menses

metabolism: sum total of processes of digestion, absorption, and the resulting release of energy

metacarpus: part of the hand between the wrist and the fingers

metatarsus: part of the foot between the tarsal bones and the toes

micturition: voiding or urinating

mitosis: cell division into two new cells, each having a complete set of chromosomes

monosaccharide: simple sugar; glucose

myelin: a lipoid substance found in the sheath around nerve fibers

myocardium: muscle of the heart

myopia: nearsightedness

nares: pertaining to the nostrils

neuron: nerve cell, including its processes

nucleus: core or center of a cell containing large quantities of DNA

occiput: pertaining to the back of the head

olfactory: pertaining to the sense of smell

ophthalmic: referring to the eyes

osmosis: passage of a fluid through a membrane

ossicle: a small bone; usually referring to the three small bones of the middle ear

palate: roof of the mouth

papilla: small nipple-shaped elevations

paralysis: loss of power of motion or sensation

parotid gland: largest of the salivary glands

patella: the kneecap

pectoral: pertaining to the chest

pelvis: any basin-shaped structure or cavity

peripheral: the outside surface or the area away from the center

pH: hydrogen ion concentration of a solution or air mixture; potential of hydrogen

phalanges: fingers and toes

pia mater: the vascular innermost covering of the brain and spinal cord

plasma: liquid part of the blood containing corpuscles

polysaccharide: complex sugar

popliteal: area behind the knee

posterior: located behind or at the back; opposite to anterior

presbyopia: farsightedness of old age due to loss of elasticity in the lens of the eye

protoplasm: living colloid material of the cell; contains proteins, lipids, inorganic salts and carbohydrates

proximal: located nearest the center of the body; point of attachment of a structure

puberty: age when the reproductive organs become functional

receptor: the sensory nerve which receives a stimulus and transmits it to the CNS

reflex: an involuntary action; automatic response

renal: pertaining to the kidney

rugae: wrinkles or folds

sagittal: longitudinal; shaped like an arrow

sartorius: muscle of the thigh

sclera: tough, white covering; part of external coat of the eye

scrotum: pouch which contains the testicles

sebum: secretion of the sebaceous glands which lubricates the skin

sella turcica: the saddle-shaped depression in the sphenoid bone

semen: male reproductive fluid containing the sperm

semilunar: half-moon shaped valve of the aorta and pulmonary artery

senescence: old age

serum: clear pale yellow fluid that separates from a clot of blood; plasma which contains no fibrinogen
sigmoid: shaped like the letter S; distal, S-shaped part of colon
sinus: a recess, cavity or hollow space
sphincter: a circular-shaped muscle, such as the anus
stapes: stirrup-shaped bone in the middle ear
sudoriferous: producing sweat
synapse: space between adjacent neurons through which an impulse is transmitted
synovia: viscid fluid present in joint cavities
systole: contraction of the ventricles, forcing blood into the aorta and pulmonary artery

tarsus: the instep
tendon: a cord of fibrous connective tissue which attaches a muscle to a bone or other structure
thrombosis: formation of a clot in a blood vessel
tibia: the larger, inner bone of the leg below the knee
turbinate: shaped like a spiral; the 3 bones situated on the lateral side of the nasal cavity
tympanum: drum; the middle ear closed externally by the ear drum

umbilicus: navel
uvula: projection hanging from the soft palate

vagina: sheathlike structure; the tube in the female extending from the uterus to the vulva
valve: a structure which permits flow of a fluid in only one direction
vein: vessel carrying blood toward the heart
ventral: front or anterior; opposite of posterior or dorsal
ventricle: small cavity or chamber as in the heart or the brain
villi: hairlike projections as in the intestinal mucous membrane
viscera: internal organs

METRIC EQUIVALENTS

Measures Commonly Used	Approximate Conversions				
Weights	Gr.	Gm.	App. oz.	Lb.	Kilos
1 grain (gr.)	1.000	0.0648	0.00208	0.0001429	0.0000648
1 gram (Gm.)	15.43	1.000	0.03215	0.002205	0.001000
1 apothecary ounce	480.	31.1	1.000	0.06855	0.0311
1 avoirdupois pound	7000.	454.	14.58	1.000	0.454
1 kilogram	15432.	1000.	32.15	2.205	1.000
Lengths	Cm.	Inches	Feet	Yards	Meters
1 centimeter	1.000	0.394	0.0328	0.01094	0.0100
1 inch	2.54	1.000	0.0833	0.0278	0.0254
1 foot	30.48	12.00	1.000	0.333	0.305
1 yard	91.4	36.00	3.000	1.000	0.914
1 meter	100.0	39.4	3.28	1.094	1.000
1 kilometer	100000.	39400.	3280.	1094.	1000.
1 mile	160903.	63360.	5280.	1760.	1609.
Volumes	Cc.	Fl. drams	Cu. in.	Fl. oz.	Quarts	Liters
1 cubic centimeter	1.000	0.270	0.0610	0.0338	0.001057	0.001000
1 fluid dram	3.70	1.000	0.226	0.1250	0.00391	0.00370
1 cubic inch	16.39	4.43	1.000	0.554	0.0173	0.01639
1 fluid ounce	29.6	8.00	1.804	1.000	0.03125	0.0296
1 quart	946.	255.	57.75	32.0	1.000	0.946
1 liter	1000.	270.	61.0	33.8	1.056	1.000

HOUSEHOLD	METRIC
1 quart	1000 ml
1 pint	500 ml
1 fluid ounce	30 ml
1 tablespoon	15 ml
1 teaspoon	5 ml
20 drops	1 ml

METRIC SYSTEM EQUIVALENTS

INDEX

A

A-V node, 41
Abdominal cavity, 2
Abdominal hernia, 35
Abdominopelvic cavity, 2
Abduction, 16
Abnormal muscular conditions, 35
Accessory organs, digestion, 79
Acne vulgaris, 103
Acromegaly, 130
ACTH, 120, 121, 125
Acute bacterial meningitis, 146
Acute kidney failure, 103
Acute nephritis, 103
Acute rheumatic heart disease, 62
Addison's disease, 130
Adduction, 16
Adenitis, 60
ADH, 98, 121
Adrenal glands, 118, 119, 125-126
    disturbances, 130
Adrenalin, 125
Adrenocorticotrophic hormone, 121, 125
Afferent nerve, 138
Air, health need, 151
Albumin in urine, test for, 98
Aldosteronism, 130
Alimentary canal, 79, 80, 83, 86
Alveolar sacs, 70
Alveoli, 67, 69, 70
Anabolism, 1
Androgens, 125
Anemia, 63
Aneurysm, 63
Angina pectoris, 62
Angular joints, 15
Anterior lobe, 120, 121
Antidiuretic hormone, 98, 121
Anuria, 103
Anvil, 142
Aorta, 45, 46, 51, 52
Aortic semilunar valve, 41, 42
Appendicular skeleton, 21
Appendix, 89
Aqueous humor, 142
Arachnoid, 135
Areola, 111
Arm, 23
Arrhythmia, 62
Arteries, 51, 52
Arterioles, 51
Arteriosclerosis, 62
Arthritis, 25, 120
Asthma, 75
Atelectasis, 75
Athlete's foot, 104
Atlas vertebrae, 22
Atrium, 41, 42
Atrial fibrillation, 62
Atrioventricular node, 41
Autonomic nervous system, 134, 138-139
Axial skeleton, 21
Axis vertebra, 22
Axons, 133, 134

B

Ball and socket joints, 15
Basal metabolism rate, 129
Bathing, health care, 152
Belly, 30
Beta cells, 127
Biceps femoris muscle, 32, 33
Biceps muscles, 32, 33
Bicuspid valve, heart, 41, 42
Bicuspids, teeth, 80, 81
Bile, 87
Bladder, 97, 98
Blood, 54-56
    composition, 54
    disorders, 63
    general circulation, 45
    lymphatic system, 59-60
    pulmonary circulation, 49-50
Blood circulatory system, 38, 41-43
Blood clotting, 55
Blood norms, 55, 56
Blood plasma, lymph, 59
Blood platelets, 55
Blood pressure, 52, 139
    disorders, 63
Blood types, 55, 56
Blood vessels, 51-53
    disorders, 62
Body, whole, 1-2
Body balance, 143
Body framework, 14-16
    bone structure, 18
    injuries and diseases of bones and joints, 25-26
    skeleton, 21-23
Body functions, coordination, 133-146
    care, 150-153
    central nervous system, 135-136
    diseases of nervous system, 145-146
    nervous system, 133-134
    peripheral and autonomic nervous systems, 138-139
    sense organs – eye and ear, 141-143
Body functions, regulators, 118-130
    adrenal glands and gonads, 125-126
    disturbances, 129-130
    endocrine system, 118-119
    pancreas, 127
    pituitary gland, 120-121
    thyroid and parathyroid glands, 123
Body movement, 28-35
    abnormal muscular conditions, 35
    attachment of muscles, 30-31
    muscular system, 28
    skeletal muscles, 32-33
Body systems, 10-11

INDEX

Boils, 104
Bones, skeletal system, 14-16
   injuries and diseases, 25-26
   skeleton, 21-23
   structure and formation, 18
   tissue, 7, 8
Bowman's capsule, 97
Brain, 1, 135-136
Brain stem, 136
Breasts, 111
   mastectomy, 115
Breathing processes, 67-75
   disorders, 74-75
   mechanics of breathing, 70, 72
   organs and structures, 69-70
   respiratory system, 67-68
Bright's disease, 103
Broken bones, 25
Bronchi, 69
Bronchiectasis, 75
Bronchioles, 69, 70
Bronchitis, 74
Bulbourethral glands, 113
Bundles of His, 41
Bunion, 25
Bursa, 15
Bursitis, 26

C

C-Gs, 125
Cancer, 151
   digestive tract, 91
   larynx, 75
   lung, 75
   reproductive system, 115
Canine teeth, 80, 81
Capillaries, 51, 70
Carbuncles, 104
Carcinoma, *see* Cancer
Cardiac arrythmia, 41
Cardiac cycle, 41-43
Cardiac muscles, 28
Cardiac sphincter, 83
Carpal bones, 23
Cartilage tissues, 7, 8
Cartilaginous joints, 15
Catabolism, 1
Cataract, 146
Cells, 1, 4-5
   tissues from, 7
Central nervous system, 134, 135-136
Cerebellum, 136
Cerebral hemorrhage, 63
Cerebral palsy, 145
Cerebrum, 136
Cervical vertebra, 15
Cervix, 111
Chemical reactions, 1
Chemical syntheses, 4
Chest, 70
Cholecystitis, 91

Chorea, 145
Choroid coat, 141
Chromatin material, 5
Chromosomes, 5, 108, 109
Chronic constipation, 91
Chronic nephritis, 103
Chyme, 83
Circulatory system, 10-11, 38-66
   blood, 38-39, 54-56
   blood vessels, 51-53
   disorders, 62-63
   general, 45-47
   health practices, 151
   heart, 41-43
   introduction to, 38-39
   lymphatic system, 59-60
   pulmonary, 49-50
Circumcision, 113
Cirrhosis, 91
Clavicles, 23
Clubfoot, 25
Coccyx, 23
Cochlea, 142
Cochlear duct, 143
Cold, common, 74
Colon, 89
Comminuted fracture, 25
Common cold, 74
Communication systems, 133
Compact bone, 18
Compound fracture, 25
Conductivity, 134
Condyloid joints, 15
Cones, of eye, 142
Congenital heart disease, 62
Congenital malformations, 25-26
Congestive heart failure, 62
Conjunctivitis, 146
Connecting neurons, 134
Connective tissue, 7, 8
Constipation, 89, 91
Convolutions, brain, 136
Convulsions, 145
Coordination of body functions, 134-146
   central nervous system, 135-136
   diseases of nervous system, 145-146
   nervous system, 133-134
   peripheral and autonomic nervous systems, 138-139
   sense organs — eye and ear, 141-143
Cornea, 141
Coronary circulation, 46
Coronary occlusion, 62
Corpus luteum, 111
Cortex, 125, 135
Corticoids, 125
Cowper's glands, 113
Cranial cavity, 1
Cranial nerves, 138
Cranium, 21
Cretinism, 129
Cushing's Syndrome, 130

INDEX

Cystitis, 103
Cytoplasm, 4, 5

D

Deltoid muscle, 32, 33
Dendrites, 133, 134
Deoxygenated blood, 49
Deoxyribonucleic acid, 5, 109
Diabetes insipidus, 130
Diabetes mellitus, 127, 130
Diaphragm, 1
    muscles, 32
Diaphysis, 18
Diarrhea, 91
Diastolic blood pressure, 52
Diet, 151-152
Digestive system, 10, 11, 79-91
    disorders, 91
    health practices, 152
    introduction to, 79-81
    large intestine, 89
    small intestine, 86-87
    stomach, 83
Diphtheria, 75
Diseases, *see* Disorders
Dislocation, 25
Disorders, of
    bones and joints, 25-26
    circulatory system, 62-63
    digestive system, 91
    excretory system, 103-104
    nervous system, 145-146
    regulators of body functions, 129-130
    reproductive system, 115
    respiratory system, 74-75
DNA, 5, 109
Dorsal cavity, 1
Ductless glands, 118
Duodenum, 83, 86
Dura mater, 135
Dwarfism, 130
Dysmenorrhea, 115

E

Ear, 142-143
    care, 153
    disorders, 146
Eating habits, 151
Eczema, 104
Effectors, 139
Efferent nerve, 138
Electric cardiac pacemaker, 41
Electron microscope, 4
Elimination of waste materials, 95-104
    disorders, 103-104
    excretory system, 95
    skin, 101
    urinary system, 97-98
Embolism, 63
Embryo, 111
Emphysema, 75, 151

Encephalitis, 145
Endocarditis, 62
Endocardium, 41, 42
Endocervicitis, 115
Endocrine glands, 118, 119
Endocrine system, 10, 118-119
    adrenal glands and gonads, 125-126
    disturbances, 129-130
    pancreas, 127
    pituitary gland, 120-121
    thyroid and parathyroid glands, 123
Endometrium, 111
Endosteum, 18
Energy, 1, 4
    oxidation process, 30, 67
Enzymes, 79
    pepsin, 83
    small intestine, 86
Epididymis, 112, 113
Epididymitis, 115
Epilepsy, 146
Epinephrine, 125
Epiphysis, 18
Epithelial tissue, 7, 8
Equilibrium, 143
Equivalents, Metric system, 160
ERV, 72
Erythrocytes, 18, 54, 56
Estrogen, 111, 126
Ethmoid sinus, 70
Eustacian tube, 142
Excretion, 89
Excretory system, 10
    disorders, 103-104
    introduction to, 95
    skin, 101
    urinary system, 97-98
Exercise, 150
Exhalation, 72
Expiratory reserve volume, 72
Extension, 16
Extensor muscles, 30, 32, 33
External respiration, 67
Eye, 141-142
    care, 153
    disorders, 146

F

Facial bones, 22
Fallopian tube, 110, 111, 115
Feet, bones, 23
    care, 150-151, 152
    flatfoot, 35
Female reproductive system, 110-112
Femur, 23
Fertilization, 109, 110
Fetus, 111
Fibers, 133
Fibrin, 55
Fibroid tumors, 115
Fibrous joints, 15

INDEX

Fibrous tissues, 8
Flat bones, 14
Flatfoot, 35
Flexion, 15
Flexors, 30, 32, 33
Follicle-stimulating hormone, 111, 121
Fontanel, 18
Food, digestion, 10, 79-81
    circulatory system, 38-39
    disorders, 91
    good eating habits, 151-152
    heart, 41-42
    in large intestine, 89
    in small intestine, 86-87
    in stomach, 83
    transport, 38
Foot, *see* Feet
Fracture, bone, 25
Framework, body, *see* Body framework
Fresh air, 151
Frontal sinus, 70
FSH, 111, 121
Fundus, 111
Furuncles, 104

G

Gallbladder, 87
Gallstones, 87, 91
Gametes, 108, 109, 110, 112
Gangrene, 62, 104
Gastric glands, 83
Gastritis, 91
Gastrocnemius muscle, 32
Gastroenteritis, 91
General circulation, 38, 45-47
GH, 121
Gigantism, 130
Gingivae, 80
Glands, 2
    adrenal, 125
    digestive, 83
    endocrine, 118
    pancreas, 127
    pituitary, 120-121
    sex, 113
    skin, 101
    thymus, 123
    thyroid and parathyroid, 123
Glaucoma, 146
Gliding joints, 15
Globin, 55
Glomerulus, 97
Glossary, 155-159
Glucocorticoids, 125
Gluteus maximus muscle, 32, 33
Gluteus medius muscle, 32, 33
Goiter, 129
Gonadotrophic hormone, 112
Gonads, 118, 119, 125, 126
    disturbances, 130
Graafian follicles, 110

Gray matter, brain, 136
    spinal cord, 136
Greenstick fracture, 25
Growth, regulation, 123
Growth hormone, 121

H

Hair, care, 152
Hammer, 142
Hand, bones, 23
Hard palate, 79
Haversian canals, 18
Head, bones, 21
Health maintenance, 150-153
Hearing function, 22, 142-143
Heart, 41-43
    disorders, 62
    health practices, 151
Heart failure, 62
Heart murmurs, 62
Heartburn, 91
Heme, 55
Hemoglobin, 54-55
Hemophilia, 63
Hemorrhoids, 63
Hepatic vein, 46
Hepatitis, 91
Heredity, 4, 5
Hernia, abdominal, 35
Herpes zoster, 104, 145
Hiatal hernia, 91
Hinge joints, 15
Hives, 104
Hormones, 118
    adrenal glands, 125
    pituitary gland, 120, 121
    reproductive system, 111, 112
Human reproduction, 108-115
    disorders, 115
    organs, 110-113
    reproductive system, 108-109
Humerus, 23
Hydrocephalus, 145
Hydrochloric acid, 83
Hypertension, 63
Hyperthyroidism, 129
Hypophyseal gland, 120, 121
Hypotension, 63
Hypothalamus, 136
Hypothyroidism, 129
Hysterectomy, 115

I

ICSH, 121
Immovable joints, 15
Impetigo contagiosa, 104
Incisors, 80, 81
Infantile paralysis, 145
Infectious causes, respiratory disorders, 74
Infectious hepatitis, 91

INDEX

Influenza, 74
Inguinal hernia, 35
Inhalation, 72
Inherited traits, 109
Injuries, bones and joints, 25-26
Innominate bones, 23
Insertion, muscle, 30
Inspiratory reserve volume, 72
Insulin, 118, 127
Integumentary system, 101
Intercostals, 32, 33
Internal respiration, 67
Interstitial cell-stimulating hormone, 121
Intervertebral discs, 22
Intestines, 95
    large, 89
    small, 86-87
Inunction, 101
Involuntary muscles, 28
Iodine, 123
Iodine tests, 129
Iris, of eye, 142
Irregular bones, 14
Irritability, 134
IRV, 72
Islets of Langerhans, 118, 127

J

Joints, 15
    injuries and diseases, 25-26

K

Kidney failure, 103
Kidney stones, 103, 130
Kidneys, 95, 97, 98
Kyphosis, 26

L

Lactogenic hormone, 121
Large intestine, 89
Laryngitis, 74
Larynx, 69
    cancer, 75
Latissimus dorsi muscle, 32, 33
Leg bones, 23
Lens, of eye, 142
Leukemia, 63
Leukocytes, 18, 55
Leukorrhea, 115
LH, 111, 121
Ligaments, 15
    injuries, 25
Liver, 87
Liver bile, 86
Lockjaw, 35
Long bones, 14
Loop of Henle, 97
Lordosis, 26
LTH, 121
Lungs, 95
    cancer, 75

Luteinizing hormone, 111, 121
Luteotropin, 121
Lymph, 59
Lymph nodes, 59-60
Lymphatic system 38, 59-60
Lymphocytes, 60
Lymphoid cells, 123

M

M-Cs, 125
Male reproductive system, 112-113
Marrow, 18
Mastectomy, 115
Maxillary sinus, 70
Measures, equivalents, 160
Medulla, 72, 125, 136
Medullary canal, 18
Meiosis, 108
Membranes, 4
Meninges, 135
Meningitis, 146
Menstruation, 111, 115
    disturbances, 130
Metabolism, 1
Metacarpal bones, 23
Metatarsal bones, 23
Metric system equivalents, 160
Microcephalus, 25
Mineralcorticoids, 125
Miner's disease, 75
Mitochondria, 4
Mitosis, 4
Mitral valve, 41, 42
Mixed nerve, 138
Molars, 80, 81
Motion, types of, 15-16
Motor neurons, 134
Motor nerve, 138
Mouth, role in digestion, 79-81
Movable joints, 15
Muscle atrophy, 35
Muscle fatigue, 30, 35
Muscle hypertrophy, 35
Muscle spasms, 35
Muscle tissue, 7, 8
Muscle tone, 30
Muscular system, 10, 28-35
    abnormal conditions, 35
    attachment of muscles, 30-31
    heart muscles, 41
    skeletal muscles, 32-33
Myocardial infarction, 62
Myocarditis, 62
Myocardium, 41
Myxedema, 129

N

Nasal cavity, 69, 70
Nasal septum, 70
Nephritis, 103
Nephrons, 97

INDEX

Nerve cells, 5, 133
Nervous system, 133-134
    central, 135-136
    disorders, 145-146
    muscles, contact, 31
    peripheral and autonomic, 138-139
    sense organs — eye and ear, 141-143
Nervous tissue, 7, 8
Neuralgia, 145
Neuritis, 145
Neurons, 133-134
Noninfectious causes, respiratory ailments, 75
Norepinephrine, 125
Nucleolus, 4
Nucleus, cell, 4, 5

O

Obesity, 151
Ophthalmologist, 153
Orchitis, 115
Organ of Corti, 141, 142, 143
Organs, 90
    body systems, 1, 10-11
Origin, muscle, 30
Ossification, 18
Osteocytes, 18
Osteomyelitis, 26
Osteoporosis, 26
Otitis media, 146
Ovaries, 118, 119, 125, 126
Ovulation, 110
Ovum, 108, 109, 110
Oxidation, 30, 67
Oxygen, circulatory system, 38-39
    heart, 41-43
Oxygenated blood, 49, 50
Oxytocin, 121

P

Pacemaker, 41
Pancreas, 118, 119, 127
    disturbances, 130
Pancreatic juice, 86
Papillae, 79
Paralysis, 35
Parasympathetic nervous system, 138-139
Parathyroid glands, 118, 119, 123
    disturbances, 129
Parotid glands, 79
Partly movable joints, 15
P.B.I., 129
Pectoralis major muscle, 32, 33
Pelvic cavity, 2
Pelvic girdle, 23
Pelvic nerve, 139
Penis, 113
Pepsin, 83
Peptic ulcers, 91
Pericarditis, 62
Pericardium, 41
Periosteum, 18

Peripheral nervous system, 134, 138-139
Peristalsis, 83
Peritonitis, 91
Pernicious anemia, 63
Personal hygiene, 152
Perspiration, 101
Phagocyte, 55
Phalanges, 23
Pharyngitis, 74
Pharynx, 69, 83
Phlebitis, 62
Pia mater, 135
Pituitary gland, 118, 119, 120-121
    disturbances, 130
Pivot joints, 15
Plasma, 54
Pleura, 70
Pleurisy, 74
Plexuses, 138
Pneumonia, 75
Poliomyelitis, 145
Polycythemia, 63
Pons, 136
Portal circulation, 46
Posterior lobe, 120, 121
Progesterone, 111, 126
Projection, 141
Prolactin, 111, 121
Pronation, 16
Prostate gland, 112, 113
Prostatectomy, 115
Prostatitis, 115
Protein-bound iodine test, 129
Protein manufacture, 4
Prothrombin, 55
Pruritus, 104
Psoriasis, 104
Ptyalin, 79
Pulmonary artery, 42, 45
Pulmonary circulation, 38, 39, 49
Pulmonary semilunar valve, 41, 42
Pupil, of eye, 142
Pyelitis, 103
Pyelonephritis, 103
Pyloric sphincter, 83
Pyloric stenosis, 91
Pyloric valve, 83

R

Radioactive iodine test, 129
Radius, arm bone, 23
Receptors, 139
Rectus abdominis muscle, 32, 33
Rectus femoris muscle, 32, 33
Red blood cells, 5
Reduction, fracture, 25
Reflex action, 136, 139
Reflex arc, 139
Regulators of body functions, 118-130
    adrenal glands and gonads, 125-126
    disorders, 129-130

INDEX

Regulators of body functions, continued
    endocrine system, 118-119
    pancreas, 127
    pituitary glands, 120-121
    thyroid and parathyroid glands, 123
Rehabilitation, muscles, 35
Renal circulation, 46
Renal pelvis, 97
Reproductive system, 10, 108-109
    disorders, 115
    organs, 110-113
Residual volume, air, 72
Respiratory system, 10, 67-68
    disorders, 74-75, 151
    mechanics of breathing, 72
    organs and structure, 69-70
Response, to stimulus, 139
Rest, 150
Retina, of eye, 142
Retroversion, 115
Rh factor, 54
Rhinitis, 75
Rickets, 25
Ringworm, 104
Rods, of eye, 142
Rotation movement, 16
Rupture, 35
RV, 72

S

S-A node, 41
Sacrospinalis muscle, 32
Sacrum, 23
St. Vitus' dance, 145
Salivary glands, 79
Salpingitis, 115
Sartorius muscle, 32, 33
Scabies, 104
Scapulae, 23
Sciatica, 145
Sclera, 141
Sclerotic coat, 141
Scoliosis, 26
Scrotum, 112, 113
Sebaceous glands, acne, 103
Sebum, 101
Sedimentation rate, erythrocytes, 56
Sella turcica, 119
Semicircular canals, 143
Seminal vesicles, 112
Seminiferous tubules, 112
Sense organs – eye and ear, 141-143
Sensory nerve, 138
Sensory neurons, 134
Sensory receptors, 141
Serratus, muscle, 32, 33
Seven-year itch, 104
Sex glands, 118, 119, 125
Shingles, 104, 145
Short bones, 14
Shoulder girdle, 23
Sickle cell anemia, 63

Sight, 141-142
Silicosis, 75
Simple fracture, 25
Sinoauricular node, 41
Sinuses, 70
Sinusitis, 74
Skeletal system, 10, 11, 14-16
Skeleton, parts of, 21-23
Skin, 95, 101
    disorders, 103-104
Skull, 21-22
Sleep, 150
Small intestine, digestion in, 86-87
Smell, sense of, 70
Smooth muscles, 28
Soft palate, 80
Somatotropin, 121
Specialized cells, 5
Sperm, 108, 109, 112, 113
Sphenoid sinus, 70
Sphincter muscles, 28
Spina bifida, 26
Spinal cavity, 1
Spinal cord, 136
Spinal nerves, 138
Spine, 22
Spleen, 38
Spongy bone, 18
Sprain, 25
Sterility, 115
Sternocleidomastoid muscle, 32, 33
Sternum, 23
STH, 121
Stiff neck, 35
Stimulus, 139
Stirrup, 142
Stomach, digestion in, 83
Stomatitis, 91
Striated muscle cells, 28
Structural units, 1-2
Subclavian veins, 59
Sudoriferous glands, 101
Sugar in urine, tests for, 98-99
Supination, 16
Sty, 146
Sweat glands, 101
Sympathetic nervous system, 138-139
Synovial fluid, 15
Systemic circulation, 38, 45-47
Systems, 10-11
Systolic blood pressure, 52, 125

T

Talipes, 25
Tarsus, 23
Teeth, 80
Tendons, 15
Terms defined, 155-159
Testes, 112, 113, 118, 119, 125, 126
    disorders, 115
Testosterone, 112, 126

INDEX

Tetanus, 35
Tetany, 130
Thalamus, 136
Thoracic cavity, 2, 70
Thrombin, 55
Thrombocytes, 55
Thromboplastin, 55
Thrombosis, 63
Thymus gland, 123
Thyroid gland, 118, 119, 123
    disturbances, 129
Thyroid stimulating hormone, 121, 123
Thyrotropin, 121
Thyroxin, 123
Thyroxin level tests, 129
Tibia, 23
Tibialis anterior muscle, 32, 33
Tidal volume, 72
Tissues, 1, 7-8
Tongue, 79
Tonsillitis, 74
Tonsils, 80
Trachea, 69
Transportation system, food and oxygen, 38-66
    blood, 38-39, 54-56
    blood vessels, 51-53
    disorders, 62-63
    general circulation, 45-47
    heart, 41-43
    lymphatic system, 59-60
    pulmonary system, 49-50
Trapezius muscle, 32, 33
Triceps muscle, 32, 33
Tricuspid valve, 41, 42
TSH, 121, 123
Tuberculosis, 26, 75
    kidneys, 103
Tubules, 97
Tumors, 26
    breast or uterus, 115
Turbinates, 70
Turner's syndrome, 130
TV, 72

U

Ulna, 23
Uremia, 103
Ureters, 97
Urethra, 97, 98
Urinary system, 97-98
Urinary tract, disorders, 103
Urticaria, 104
Uterus, 111
    disorders, 115
    removal, 115
Uvula, 80

V

Vagina, 111
Vagus nerve, 138
Varicose veins, 62
Vas deferens, 112, 113
Vascular tissues, 8
Vasopressin, 121
Vastus lateralis muscle, 32, 33
Veins, 51
Vena cava, 42, 45, 46, 51
Ventilation, 151
Ventral cavity, 1
Venules, 51
Vertebrae, 22
Vertebral column, 22
Villi, 86
Virilism, 130
Viruses, respiratory disorders, 74-75
Visceral organs, 138
Vital capacity, breathing, 72
Vitamin K, 55
Vitreous body, 142
Voluntary muscles, 28

W

Waste materials, elimination, 89, 95-104
    disorders, 103-104
    excretory system, 95
    skin, 101
    urinary system, 97-98
Water balance, 98
Webbed fingers and toes, 26
White matter, brain, 136
    spinal cord, 136

Z

Zygote, 109, 111